Fundamentals of Construction Law

Printed in the United States of America.

05 04 03 5 4 3

Fundamentals of construction law / Fred D. Wilshusen, Carina Y. Enhada, Cheri Turnage Gatlin (eds.)
p.cm
Includes index.
ISBN 1-57073-966-8
1. Construction contracts—United States. 2. Construction industry—Law and legislation—United States. I. Wilshusen, Fred D. 1957-II. Enhada, Carina Y. III. Gatlin, Cheri Turnage.

KF902.F86 2001
343.73'278624—dce21

2001046059

Contents

About the Editors

CARINA Y. ENHADA

Carina Y. Enhada is Senior Counsel for Dillingham Construction Corporation, based in Pleasanton, California, where she provides legal advice on all matters for Dillingham and its subsidiary companies throughout the United States.

Prior to joining Dillingham, Ms. Enhada practiced in Hawaii and Washington, where her practice concentrated in representation of institutional owners, general contractors, sureties, and subcontractors in various construction related matters including contract drafting and review, project management and administration, insurance and risk management matters, construction defect issues, lien and bond claims and other construction related litigation, arbitration and mediation. Ms. Enhada was most recently "Of Counsel" with the law firm of Short, Cressman & Burgess, PLLC of Seattle, Washington.

Ms. Enhada is a member of the Hawaii and Washington Bar Associations. She is also a member of the American Bar Association, including the ABA Forum on the Construction Industry where she has been serving as a steering committee member of Division 5, Contract Negotiations, Performance Administration, for the last three years. Ms. Enhada is a graduate of the University of San Francisco School of Law where she served as a member of the Law Review.

CHERI TURNAGE GATLIN

Cheri Turnage Gatlin is a partner in the Jackson, Mississippi law firm of Ott & Purdy, P.A. Her practice is concentrated in the areas of construction litigation, mediation, and arbitration. Her clients include general contractors, subcontractors, owners, and sureties.

Ms. Gatlin is an active member of the American Bar Association and of American Inns of Court. She is a member of the ABA Forum on the Construction Industry and the Fidelity and Surety Law Committee. She currently serves on the Steering Committee for Division Five of the Forum, and is a member of the Mississippi Bar Association, the Mississippi Young Lawyers Division, and the Hinds County Bar Association. She serves as ABA District Representative for Mississippi and is co-chair of the ABA Young Lawyers Division Business Law Subcommittee. An active alumnus of Mississippi College School of Law, she serves as coach for the law school's Moot Court Competition Team for the Duberstein National Moot Court Competition.

She is co-author of the following publications: *Annual Seminar*, Memphis Bar Association Bankruptcy Section, 1996; *Discovery of Information From An Insurer*, 3 Fidelity Law Association Journal 71, 1997; and *Analyzing Claims for Lost Labor Productivity*, The Construction Lawyer, January 2000.

FRED D. WILSHUSEN

Fred D. Wilshusen is a partner in the Law Firm of Thomas, Feldman & Wilshusen, L.L.P., a law firm that concentrates on construction law and related litigation representing all segments of the construction industry. Mr. Wilshusen received his undergraduate degree from Austin College and his J.D. From Texas Tech University in 1983.

Mr. Wilshusen is the co-author of: "How to beat a 'No Damage for Delay' Clause," 9 THE CONSTRUCTION LAWYER 17 January, 1989; "Construction Law: Annual Survey of

Texas Law," 44 S.W. L.J. 2101 (1990); Texas Lien and Bond Claims Handbook (4th ed. 2000). He is also author of several construction law related articles for various publications. He was the 1998 Chair of the Dallas Bar Construction Law Section and is the Chair of the Steering Committee of Division 5 of the Forum Committee on the Construction Industry of the American Bar Association.

About the Authors

LYNN R. AXELROTH

Lynn R. Axelroth is Managing Partner of the Philadelphia office of Ballard Spahr Andrews & Ingersoll, LLP, and is a partner in the firm's Real Estate Department. Her practice includes representing corporate, institutional, individual and public owners, public/private joint venturers, prime and trade contractors, design professionals and lenders, with respect to complex development projects, construction and design contracts, financing and alternative dispute resolution. She was the first Chair of the Owners and Lenders Division of the ABA Forum on the Construction Industry and now serves on the Forum's Governing Committee, is Co-Chair of the ABA Real Property Section's Committee on Construction Lending, a member of the American College of Real Estate Lawyers, a fellow of the American College of Construction Lawyers, and an arbitrator for the American Arbitration Association. She lectures and publishes extensively on construction law and real estate development and finance. She graduated *cum laude* from The Law School of the University of Pennsylvania, where she was an Editor of The Law Review, and summa cum laude from Temple University, where she was Phi Beta Kappa and a President's Scholar.

ERIC A. BERG

Eric A. Berg is an attorney at the Chicago office of Piper Marbury Rudnick & Wolfe. Mr. Berg focuses his practice on construction law, in both litigation and transactional work. As a litigator, Mr. Berg has represented owners, contractors, subcontractors, architects, engineers, and lenders. Specific lawsuits include defending the municipal owner of a water treatment facility against a contractor's $1.4 million claim, representing a general contractor joint venture regarding minority-owned business enterprise participation levels, and obtaining a not-guilty verdict on behalf of a sports stadium designer in a $500,000 personal injury claim. Mr. Berg also has extensive experience in other areas of commercial litigation.

Mr. Berg's transactional work includes drafting and negotiating contracts for the design, construction, rehabilitation, and construction management of hotels, hospitals, office buildings, apartment buildings, private residences, retail stores, and warehouses. Mr. Berg is well-versed in all facets of the Americans with Disabilities Act and the Fair Housing Act. For numerous clients Mr. Berg has provided counsel on interpretation of and compliance with the ADA, the FHA, and other federal, state, and local accessibility statutes.

GREGORY L. CASHION

Gregory L. Cashion is a partner in the Nashville law firm of Smith & Cashion, PLC, where he devotes his practice to construction law. Mr. Cashion earned his B.S. in Civil Engineering and J.D. degrees from the University of Tennessee. He has lectured before the Tennessee Bar Association, the Nashville Bar Association, the Association of General Contractors, the American Builders and Contractors Association, and the Tennessee Road Builders Association on construction law. He co-wrote *Fifty-State Construction Lien and Bond Law*, Chapter 43, Wiley Law Publications, Tennessee, 1992. Mr. Cashion wrote the article "Checklist for Enforcing a Miller Act Claim," *Construction*

Lawyer, Volume 16, No. 4, October 1996, published by the American Bar Association Forum on the Construction Industry. Mr. Cashion is a member of the Nashville, Tennessee and American Bar Associations, the Tennessee Society of Professional Engineers, and the National Society of Professional Engineers.

JOHN T. CLAPPISON

John T. Clappison is a member of Clark Hill, PLC, in Detroit, Michigan. He practices primarily in the areas of construction law and professional liability, representing public and private owners, general contractors, design professionals, and subcontractors/suppliers.

Mr. Clappison received a B.S. in Civil Engineering, with a concentration in Construction Management, from the University of Michigan in 1987. He graduated from the University of Texas Law School in 1993. Prior to law school, Mr. Clappison worked as an engineer for a general contractor on a variety of engineering and heavy/highway construction projects.

J. COBBIE DE GRAFT

Mr. de Graft is an attorney who specializes in government contracts, construction law, and technology law. He practices before various federal forums including the General Accounting Office, the Court of Federal Claims, and the U.S. District Court. He also manages Federal Access (www.fedaccess.com), a web-based resource center for small business government contractors. He is admitted to the bar in the District of Columbia and New York, and is a graduate of the Notre Dame Law School.

C. ALLEN GIBSON, JR.

C. Allen Gibson, Jr. is a principal in the Charleston, South Carolina law firm of Buist, Moore, Smythe & McGee, P.A.

He received his B.A. degree with high distinction in 1979 and his J.D. degree in 1982 from the University of Virginia where he was Order of the Coif. His primary area of practice involves the representation of parties in the construction industry including contract negotiation, arbitration and litigation of construction, and surety law matters. Mr. Gibson is the Chair of the American Bar Association's Forum on the Construction Industry and is a Fellow in the American College of Construction Lawyers. He is on the Panel of Arbitrators and is a Mediator for the American Arbitration Association, has served as an arbitrator and mediator in numerous cases, and has been a Special Municipal Judge for the City of Goose Creek, South Carolina. He is also a member of the National Construction Dispute Resolution Committee. Mr. Gibson is a member of the Fidelity and Surety Committee of the Torts and Insurance Practice Section of the ABA, and has served as a vice-chair of the Fidelity and Surety Committee and a member of the Construction Law Committee of the International Association of Defense Counsel. Mr. Gibson has served on the faculty of the IADC Defense Counsel Trial Academy in Boulder, Colorado. Mr. Gibson is a past president of the Young Lawyers Division of the South Carolina Bar, has served on the South Carolina Bar Board of Governors, and is presently a member of its House of Delegates. He is also the head of the Construction Practice Group at Buist, Moore, Smythe & McGee, P.A. and is a frequent lecturer on construction topics. He represents architects, owners, general contractors, subcontractors, suppliers and sureties in negotiation, arbitration, and litigation.

CHARLES R. GOSSAGE

Mr. Gossage is a Senior Associate in Thelen Reid & Priest's Los Angeles office. A Member of Thelen's Construction Group, Mr. Gossage provides a full range of legal services to owners, contractors, engineers, developers, and sureties in connection with public and private construction. Mr. Gossage has prosecuted and defended construction claims in state and federal courts, as well as bankruptcy court. Mr. Gossage received his Bachelor of Science from Arizona State University and his law degree from Pepperdine University School of Law.

ALAN E. HARRIS

Alan E. Harris is a partner with the San Francisco, California firm of Farella Braun & Martel LLP. Mr. Harris has an extensive background in construction law derived from his 30 years of experience. His litigation experience includes the preparation and trial of complex construction cases involving defects and delay damages. He has reviewed hundreds of construction contracts for clients and has acted as their legal advisor to resolve construction issues before they became claims or disputes. He has mediated dozens of disputes both for his clients and as the neutral party chosen by the claimants. He has also acted as an arbitrator, court-appointed special master, and expert in a large number of construction cases.

Mr. Harris received an A.B. in chemistry from the University of California at Berkeley and his J.D. from Boalt Hall School of Law, University of California at Berkeley. He was admitted to the State Bar of California in 1970 and became a partner in the firm in 1976. He formerly served as Chair of the firm's Executive Committee and as Chair of its Litigation Department.

Mr. Harris is a Fellow and member of the Board of Governors of the American College of Construction Lawyers. Mr. Harris is Chair of the Bar Association of San Francisco's Client Relations Committee and a member of the American Bar Association's Forum Committee on the Construction Industry.

Mr. Harris has frequently addressed the American Bar Association, the Construction Superconference, the Practicing Law Institute, and other organizations on the subject of construction claims and defenses. He has been a regular lecturer at Boalt Hall on these subjects.

L. WEAREN HUGHES

L. Wearen Hughes is a member in the Nashville office of the law firm of Bass, Berry & Sims PLC, where he practices in the areas of commercial litigation (including construction and environmental litigation), construction law, and bankruptcy/creditors' rights. He received his B.A. degree, cum laude, from Vanderbilt University. He received his J.D. degree from Vanderbilt University School of Law in 1997 where he was Order of the Coif. Mr. Hughes is a member of the Nashville and American Bar Associations. In the American Bar Association, he is a member of the Litigation Section, the Litigation Section's Committee on Construction Litigation, and the Forum on the Construction Industry. He is on the editorial staff of *Construct!*, the publication of the Committee on Construction Litigation, for which he writes articles on recent developments in Tennessee construction law. He is a frequent speaker on construction law. He also is on the Panel of Neutrals for the American Arbitration Association and is a Rule 31 Mediator for General Civil Mediations in Tennessee.

THOMAS M. KERNANEN

Thomas M. Keranen is a principal in the Michigan law firm of Federlein & Keranen, P.C. Mr. Keranen received his BS Civil Engineering from Michigan Technological University in 1972, and received his MS Engineering from the University of Michigan in 1976. In 1981, Mr. Keranen received his JD from the Detroit College of Law.

Mr. Keranen is a Professional Engineer in the State of Michigan, and prior to entering into the legal profession, spent a number of years in design, construction, and project administration on a variety of projects ranging from residential, commercial, interstate highways, airports, educational facilities, energy and communications facilities, municipal works, and the like.

Mr. Keranen is admitted to the bar in the states of Michigan, Wisconsin and Florida, the United States Supreme Court, the United States 6th Circuit Court of Appeals, various United States District Courts and the United States Court of Federal Claims. Mr. Keranen is an active participant in the American Arbitration Association, and frequently serves as a mediator or arbitrator.

Mr. Keranen maintains an active practice in construction law, representing all parties to the construction process.

Mr. Keranen is a frequent presenter and author on legal topics pertaining to the construction industry, and has served as adjunct faculty at Lawrence Technological University, School of Architecture, since 1991.

BRIAN P. KIRWIN

Brian P. Kirwin is a Partner with the Winter Park, FL, law firm of Lyon Kirwin Norris, P.A. Mr. Kirwin concentrates his practice in all aspects of construction law from contract drafting to claims resolution and litigation. His clients include general contractors, school boards, counties, municipalities, and water districts, among others, and he has represented contractors and owners in the full range of construction legal matters including preparation and negotiation of complex construction contracts, bidding and procurement disputes, payment and performance bond claims, design and plan errors, and defense of delay, disruption, acceleration, inefficiency, and lost labor productivity claims. Mr. Kirwin received both his undergraduate degree and law degree from the Catholic University of America, in Washington, D.C., where he served as an Editorial Assistant on The Catholic University Law Review. He is a member of the ABA Forum Committee on the Construction Industry and the Florida Casenote Editor of the American Bar Association Litigation Section's *Construct!* magazine.

ANDREW D. NESS

Andrew D. Ness is the managing partner of the Washington, D.C. office of Thelen Reid & Priest LLP. Mr. Ness' practice focuses on dispute resolution and counseling with respect to construction and design-related matters representing primarily owners/developers, engineers, and prime contractors. Mr. Ness has been lead counsel on a wide variety of major construction disputes in federal and state courts, and also has extensive experience in the use of mediation and other alternative dispute resolution techniques to resolve construction disputes. Mr. Ness is the author of numerous articles on construction law topics and speaks frequently on a wide range of construction law issues. He received a Bachelor's degree in electrical engineering, with distinction, from Stanford University, and graduated *magna cum laude* from Harvard Law School in 1977, where he served as articles editor of the *Harvard Journal on Legislation*. After law school, he served as law clerk to the Chief Judge of the U.S. District Court for the Northern District of California before commencing his construction law practice.

SHERRY A. NOHARA

Sherry A. Nohara is an associate in the law firm of Oliver, Lau, Lawhn, Ogawa & Nakamura and practices in the areas of construction law, surety and fidelity law, and general commercial litigation. Ms. Nohara received her law degree from Hastings College of Law in 1996.

EDWARD NEAL POLLARD

Mr. Edward Pollard is an attorney with the North Carolina firm of Conner Gwyn Schenck PLLC. Mr. Pollard received his B.S. in Civil Engineering from North Carolina State University in 1982 and his J.D. from Wake Forest University. Prior to law school Mr. Pollard worked as Civil Engineer with the Charleston Naval Shipyard in South Carolina. He held that position from 1982 to 1987, when his job became structural engineer with the Southern Division, Naval Facilities Engineering Command, Charleston, South Carolina. After law school, from 1993 to 1996 Mr. Pollard served as judicial clerk to the Honorable Hiram H. Ward, United States District Court Judge, in Winston Salem, North Carolina. From 1996 to 1999 he worked as a construction lawyer at Bugg & Wolf in Durham, North Carolina.

ROBERT A. RUBIN

Robert A. Rubin is a partner in the law firm of Postner & Rubin, New York, New York. His practice is limited to construction matters, principally the resolution of complex construction disputes. In addition to representing clients, Rubin has served as a mediator, arbitrator, special master, neutral adviser, and as a member of dispute review boards. He is a member of the New York State Bar and a Professional Engineer in New York. Rubin is a past president of the American College of Construction Lawyers, a Fellow of the American Society of Civil Engineers and a member of the Moles. He has spoken and written extensively on construction law-related topics. Recent publications (as author or co-author) include: *New York Construction Law Manual*, West Group (1992; 1999 Cumulative Supp.); *Construction Claims: Prevention and Resolution*, 3rd Ed, John Wiley & Sons (1999); New York chapter of the *State-by-State Guide to Architect, Engineer and Contractor Licensing*, Aspen Publishing (1999); and "Scope of Work Disputes," *Construction Litigation*, 2nd Ed, Practising Law Institute (1993).

CHARLES F. SEEMANN, JR.

Mr. Seemann is a partner in the firm. He joined DK&S in 1965 after three years of active service in the United States Navy. He was a director of the New Orleans Association of

Defense Counsel, and of the New Orleans Legal Assistance Corporation, and served as president of both organizations.

Mr. Seemann has been a speaker on construction law topics at continuing legal education programs for members of the Louisiana and Florida Bar Associations, and has been a featured speaker on construction and contract surety law topics before a wide variety of construction industry groups, including national meetings of the ASME and ASCE, the New Orleans Chapter of ASHRAE, Local Chapter of AGC, and the Greater New Orleans Business Roundtable, and at meetings of the Louisiana Association of School Executives and the National Institute of Municipal Law Officers. He has also presented and published papers on fidelity bonds at meetings of the American Bar Association.

Mr. Seemann is also a member of the American Arbitration Association's panel of arbitrators for construction industry disputes, and has had AAA's arbitration training. He has also had AAA's mediator training for construction industry disputes, and is on AAA's mediation panel as well.

Mr. Seeman is a graduate of Tulane University where he received an L.L.B. in 1962 and a B.S. in geology in 1959.

DAVID A. SENTER

David A. Senter is a member of the Greensboro, North Carolina firm of Adams Kleemeier Hagan Hannah & Fouts, PLLC, where he is the chair of the firm's Construction and Surety Practice Group. Since joining the firm in 1984, Mr. Senter has concentrated his practice in the areas of construction and surety, commercial litigation, and commercial collections. He has made various presentations on construction and collections related topics to national and local trade and professional associations. Mr. Senter has also authored numerous construction and collections related articles in bar and trade publications. He is also the North Carolina and Fourth Circuit Casenote writer for *Construct!*, the publication of the Construction Litigation Committee of the ABA's

Litigation Section, and is a construction law writer for the *Campbell Law Observer.*

Mr. Senter is presently the Chair of Division 9, Specialty Trade Contractors and Suppliers, of the ABA Forum on the Construction Industry and is a past member of the council of the North Carolina Bar Association's Construction Law Section. Mr. Senter is a member of the Forum on the Construction Industry, the Tort and Insurance Practice Section, and the Litigation Section of the American Bar Association. Mr. Senter is a member of the panel of arbitrators of the American Arbitration Association.

Mr. Senter received his B.S. degree, *cum laude*, in 1981, and his J.D. degree, *cum laude*, in 1984, both from Wake Forest University.

RICHARD F. SMITH

Richard F. Smith is Of Counsel to the Vienna, Virginia Law firm of Smith, Pachter, McWhorter & Allen, P.L.C. He received his B.S. degree from Wake Forest University, his LL.B. degree from the University of Virginia and his LL.M. degree from George Washington University. For over 30 years Mr. Smith has been primarily engaged in matters relating to construction and contract disputes. More than 90 percent of his practice involves dispute resolution in construction related issues on behalf of contractors, owners, and subcontractors, including public and private projects. He is on the AAA Panel of Arbitrators and has a successful mediation practice. He is on the Governing Committee of the ABA Forum on the Construction Industry and is a member of the American College of Construction Lawyers. His experience includes disputes dealing with specification interpretation, CPM scheduling, delay evaluation, differing site conditions, defective specifications, change order pricing, inefficiency claims, and contract administration issues. He has written many articles relating to construction issues and often speaks on construction related issues throughout the country.

LINDA M. THOMAS-MOBLEY, PHD, JD

Dr. Thomas-Mobley holds a faculty appointment in the Building Construction Program of the College of Architecture at the Georgia Institute of Technology. She holds a Bachelor's and Master's degree in Civil Engineering from the University of Florida, a Juris Doctor from the University of Miami, and a Doctor of Philosophy in Architecture from Georgia Tech. At Georgia Tech, Dr. Thomas-Mobley teaches Construction Law, Construction Contracting, Construction Estimating, Construction Cost Management, Safety and Environmental Issues for Facility Managers, and Construction Industry Issues and Initiatives. Dr. Thomas-Mobley served as a Civil Engineer Corps officer in the U. S. Navy, and was the director of temporary construction for the 1996 Olympic Village in Atlanta, Georgia. She joined the Building Construction Program as Research Coordinator for the Construction Resource Center in 1997 after working as a project manager in private industry and attorney at law in civil litigation. Dr. Thomas-Mobley is a member of the Georgia Bar. In 1998 she was the recipient of the Outstanding Faculty for Education and Practice Award in the College of Architecture.

JAMES P. WAGNER

Jim Wagner is a shareholder in the law firm of Stafford Frey Cooper in Seattle, Washington. Mr. Wagner received his undergraduate degree at Washington State University, and is a 1979 graduate of Stanford Law School. He is admitted to practice in both Washington and Alaska. He has been a member of the Forum on the Construction Industry since 1980 and has served on several division steering committees over the years. His practice focuses on construction and other commercial litigation, including insurance coverage litigation. In recent years an increasing percent of his case load has involved defense of developers, contractors, and design professionals in water intrusion and other types of construction defect cases.

KEITH Y. YAMADA

Keith Y. Yamada is a principal in the law firm of Oliver, Lau, Lawhn, Ogawa & Nakamura, and has practiced in Honolulu for 16 years. His firm represents many general contractors, subcontractors, owners, and sureties in construction-related disputes involving federal, state, and private projects. He has practiced before arbitration panels, administrative agencies, and federal and state courts, and has also served as an arbitrator and mediator in commercial cases.

Special Acknowledgments

Chapter 1 *The Owner's Perspective* Lynn Axelroth

Special thanks to Thomas J. Ragonetti, Otten, Johnson, Robinson, Neff & Ragonetti, P.C., Denver, Colorado for use of his previously prepared paper on the Owner's Perspective for the American Bar Association, Forum on the Construction Industry.

Chapter 2 *The Design Professional* Thomas M. Keranen

Special thanks to Danny B. O'Conner, Severn & O'Connor, Frederick, Maryland for use of his previously prepared paper on the rights and responsibilities of the architect for the American Bar Association, Forum on the Construction Industry.

Chapter 3 *Role of the Prime Contractor* John T. Clappison

Special thanks to Charles F. Seeman, Jr., Deutsch, Kerrigan, & Stiles, New Orleans, Louisiana for his previously prepared paper on general contractors for the American Bar Association, Forum on the Construction Industry.

Chapter 4 *The Role of the Subcontractor* David A. Senter

Special thanks to Steven M. Siegfried, Siegfried, Kipnis, Rivera, Lerner & Delatorre, P.A., Coral Gables, Florida for his previously prepared paper on construction subcontracting for the American Bar Association, Forum on the Construction Industry.

Chapter 5 *A Primer in Contract Surety Law* Charles R. Gossage
Special thanks to John. L. Spiegel, Plowman, Spiegel & Lewis, P.C., Pittsburgh, Pennsylvania, for his previously prepared article on contract surety law for the American Bar Association, Forum on the Construction Industry.

Chapter 8 *Scheduling for a Construction Project* Richard F. Smith
Special thanks C. Allen Gibson, Jr., Brist, Moone, Smythe, & McGee, P.A., Charleston, South Carolina for his previously prepared article on construction scheduling for the American Bar Association, Forum on the Construction Industry.

Chapter 10 *Default Termination* Gregory L. Cashion
Special thanks to Charles F. Seemann, Jr. of Deutsch, Kerrigan & Stiles, New Orleans, Louisiana for use of his previously prepared paper on default termination for the American Bar Association, Forum on the Construction Industry.

Chapter 12 *Construction Insurance: An Introduction* James P. Wagner
Special thanks to F. Malcolm Cunningham, Jr., Cunningham & Self, P.A., West Palm Beach, Florida, for use of his previously prepared article on construction insurance for the American Bar Association, Forum on the Construction Industry.

Chapter 13 *Alternative Dispute Resolution in the 21st Century* Alan E. Harris
Special thanks to Leslie King O'Neal, Holland & Knight, Orlando, Florida, for her previously prepared article on alternative dispute resolution for the American Bar Association, Forum on the Construction Industry.

Introduction

As two past chairs of the Division 5 Steering Committee of the American Bar Association's Forum on the Construction Industry, it is with considerable pleasure that we introduce the Forum's new book on the fundamentals of construction law.

This book springs from a program produced by Division 5 some years ago. That program was based on the concept that what sets construction lawyers apart from other lawyers is not just an understanding of the principles of law which govern construction work, but also their practical understanding of the industry itself and how it works. For the most part, construction contracting is a consensual process in which the parties negotiate the best deal they can, and then live with the consequences of that deal. It is therefore important for construction lawyers to be knowledgeable about both the legal and practical aspects of the construction industry. This knowledge cannot be fully acquired overnight, and it is fully understood only with experience.

Many years ago, Division 5 recognized that it would give lawyers new to the construction practice a big head start to get a basic explanation of the way the industry works, and the economic imperatives faced by each of the parties in various circumstances. Our program was directed to that purpose, and this book completes that process.

An understanding of the basics regarding the construction industry is critical to giving proper advice to a construction client in almost any situation. To help a client negotiate and draft a contract, one needs to understand the economic needs, not only of the client, but also of the other party to the contract. Only with this understanding will the lawyer be able to negotiate the best deal for the client without losing the deal by overreaching.

When a dispute develops, it is even more important for the lawyer to know where to look in the contract for the pertinent language, and to have some sense of whether your client has any economic leverage in the situation presented. Finally, when a construction dispute must be litigated or arbitrated, it is critical not only to understand the strengths and weaknesses of your client's case, but also to understand enough about the process and its details to be able to examine and cross-examine witnesses correctly.

Gaining a full understanding of these complex issues is a task which probably cannot be achieved just by reading this single volume. The authors of this book, however, have given the reader a very good start. This book will give any newcomer to the construction practice a strong foundation for understanding the thought process of the construction industry and the basics of the legal principles which apply to it. It will also help those who have been practicing for some time to develop a greater insight into the industry and clients they serve.

Charles F. Seeman, Jr.
C. Allen Gibson, Jr.

1 The Owner's Perspective

Lynn Axelroth

I. INTRODUCTION

A construction project is much more than an assemblage of bricks and mortar. It is a complicated series of relationships between the parties involved in the project. When the relationships are functioning properly, the result is a finished product meeting the needs of both the end-user and the people creating that product. Unfortunately, in many projects, that is not what happens. Even more problematic is the fact that of all of the participants in the construction process, the owner often is the least knowledgeable and the most responsible.

This chapter is intended to address some of the most significant concerns of owners and a number of ways to address those concerns. Because construction lenders often become unwilling owners in a failing project, this chapter also highlights issues that construction lenders should consider.

A. The Owner's and Lender's Objectives

An owner generally wants to achieve three things, although the relative importance to the owner of each goal may vary from project to project. First, an owner has a specific scope of work in mind, sometimes known as the "program." Second, the owner needs the work to be completed within a particular time frame, usually described in a project schedule. Third, the project must be completed within the owner's budget. In slightly more specific terms, the owner would like the project to be constructed in a "good and workmanlike" manner, that is, using good quality materials and generally accepted construction techniques.

Finally, it is essential to the owner that the project, at completion, be free of mechanics' liens and claims. Not only will mechanics' liens jeopardize the owner's title to the project in a legal sense, but payment of them over and above the agreed upon price will create additional costs for the project, over and above the owner's budget.

For the most part, any lender supplying financing for a construction project will have the same general objectives as the owner. But because of the relationship a construction lender typically has to a construction project, the lender's general objectives go beyond the owner's in several respects.

In the typical construction project, a construction lender provides most, if not all, of the funding for construction. The lender may disburse payments to the owner for payment to the contractor and other parties, or the lender may disburse payments directly to the contractor as construction progresses. Although the project's owner may be liable for repayment of the construction debt (*i.e.*, a recourse loan), the lender's primary collateral and source of repayment in the event of a default by the owner is the lender's interest in the project. Accordingly, the lender will typically require a first lien position on the project. Two of the lender's key additional objectives, then, are to maintain that first lien position under all circumstances and to be able to step easily into the owner's shoes in the event of a default by the owner, in order to protect and exploit the project's collateral value. The lender needs to be satisfied that no funds are disbursed for work that was not performed, or performed improperly, or for materials which are not owned or not in accordance with the architect's plans and specifications.

In most projects, the construction loan represents but a portion of the ultimate value of the project—say, typically somewhere between 50% and 90% of the project's ultimate value. The lender also will want to make sure that it maintains the loan to value ratio under which the lender's loan originally was underwritten. That means the project must be completed for the projected budgeted amount within the time projected.

Although the lender and the owner may have different priorities from time to time, both will focus on the scope, time and cost of the project. The method of achieving the desired results, however, varies depending upon a number of circumstances.

B. Contract Documentation and Form Contracts

Before addressing those circumstances, it is important to understand a little bit about form contracts. The construction industry, perhaps more than any other field, abounds with pre-printed form contracts. In private projects, the most widely used form contracts are published by the American Institute of Architects (AIA). Other construction industry groups, such as the Associated General Contractors of America (AGC), the Design Build Institute of America (DBIA) and the Engineers Joint Contract Documents Committee (EJCDC), a venture between the National Society of Professional Engineers and the Consulting Engineers Council and the American Society of Civil Engineers, have developed standard forms to cover almost every typical construction relationship. Other entities produce forms as well, staking claims to very specialized segments of the industry.

It is difficult to imagine a preprinted form which could respond to all of the peculiar complexities of any given construction project without substantial modification. Moreover, each of the industry forms bears the bias of its publisher. For example, the AIA is a trade association whose primary mission is the advancement of architects. Similarly, the AGC has as its primary duty the promotion of the interests of contractors. Regardless of what these entities state in their respective promotional materials, their agreements are not neutral and generally do not benefit owners or lenders.

Forms should be viewed as a starting place only, and must be carefully reviewed and changed to meet the needs of the project and the parties. There is no such thing as "boilerplate" language. The practitioner must read and understand every provision. If the project is particularly complex or costly, it may be worthwhile to seek advice from a lawyer experienced in the industry to produce a custom contract for that project. The balancing act performed by most lawyers is to determine whether the transaction is sufficiently close to any one of the form documents so that a form can be used as a starting point and can be modified as necessary to fit the particulars of the transaction, or whether the transaction is so unique that the most efficient and effective way to document it will be to start from scratch, perhaps referring to form documentation for guidance on how to treat some standard and recurring construction issues.

In many cases, the choice will not be left to a lawyer; the lawyer simply will be given a set of already executed form documents to deal with in addressing or resolving a construction controversy. In fact, unlike many other transactional settings where lawyers play a central role in documenting the transaction at the outset, in the construction context many parties believe that it is proper, or even preferable, for the principal parties to attempt form documentation on their own, without a lawyer's advice at the time.

A final note: construction projects are multi-party transactions, but rarely is it the case that all or most of the parties involved in the project will be parties to the same document or documents. In fact, most construction transactions are documented in a series of two-party contracts, such as owner/architect, owner/contractor, and contractor/subcontractor. Nevertheless, the conduct of most construction projects contemplates a complex set of interrelationships, and respective rights and obligations, which cannot be adequately described and addressed in a series of purely two-party documents. This problem raises a number of drafting issues.

First, the rights and obligations of each party with respect to each of the other parties usually must be addressed if the project is to be carried out successfully. This objective may be achieved by the use of special provisions in the two-party documents which relate them one to another; it is more often the case, however, that those two-party documents refer to and incorporate a set of general understandings, set forth in some separate document, under which all of the parties are jointly operating. In the case of the AIA documentation, that document is the "General Conditions of the Contract for Construction" (AIA A201 and variations thereof), which the owner/architect agreement, the owner/contractor agreement and the contractor/subcontractor agreements, among others, will refer to and adopt by incorporation. Please keep in mind that to the extent one modifies the standard general conditions, all of the contract documents referring to them must reference the same, now modified, version of them.

Second, to the extent various two-party documents address the same subject—such as dispute resolution—it is important that they be consistent in approach with one another, unless there has been a conscious decision made to treat a particular subject differently in one or more of the agreements. Even in that event, it is important to be sure that the variation in treatment in one document (and the corresponding two-party relationship) does not create problems in the other agreements.

II. THE RELATIONSHIP BETWEEN THE OWNER AND THE ARCHITECT

A. Introduction

The chapter on Delivery Systems identifies many ways in which a construction project may begin and be prosecuted. It will be helpful in understanding the issues that follow to read that chapter first. However, a very brief overview follows.

1. Contracting Structures

Although there are many variations, there are three basic contracting structures found in the construction industry. The traditional structure is for the owner to engage an architect to design the work, and then engage a general contractor to perform the work, aided by various subcontractors employed by the general contractor to complete the work. This structure places the requirement of coordinating, managing and performing the work on the general contractor. The architect may perform certain construction administration services, but the architect is not responsible for how its design is constructed (sometimes referred to as "the means and methods"). The owner has privity of contract with the general contractor and the architect, but not with the trade contractors or subcontractors. Owners often view this structure as having the least risk, but it can be expensive because the contractor is adding its overhead and profit on top of those of its subcontractors.

Another structure is one in which the owner again engages an architect to design the work and also employs a construction manager. The owner will then enter into multiple construction contracts directly with the trade contractors, or the construction manager will enter into such contracts, but strictly as the agent of the owner. The construction manager coordinates the work of the trade contractors and performs scheduling, but it is not in direct privity with the trade contractors. Therefore, unlike a general contractor, it generally is not responsible for the failure of the work to comply with the requirements of the plans and specifications, for cost overruns or for failure to complete the work within a specific period of time. The construction manager receives a fee for its services, and the owner is responsible for payment of the trade contractors. This structure can be less expensive for an owner because it eliminates the general contractor's addition of its overhead and profit to that of the subcontractors. However, it shifts the risks of cost overruns and failure to meet the required completion date to the owner.

A variation of this structure is an arrangement where the construction manager enters into the trade contracts directly, and not as an agent of the owner. In that case, the construction manager is responsible for the work. In reality, although the contract may be referred to as "construction management," it essentially is a general contract with a slightly different compensation arrangement.

Another basic contract structure available is design/build. With this method, the responsibility for the design and construction of the project is combined in and contractually delegated to one entity. That entity, like a general contractor, hires the trade subcontractors; but unlike a general contractor, it designs the project as well. This method of contracting is advantageous because it avoids the common problem of the architect and the contractor pointing fingers at each other if there is a problem with the project. However, it deprives the owner of an independent party (the architect) to keep an eye on the contractor. Also, until the project is designed, an accurate guaranteed maximum price or lump sum cannot be established.

As noted above, in many instances the construction process will begin with the owner selecting a design professional to help assess the owner's needs and define the project. While the design professional may be an engineer or an architect, depending upon the project, for ease of reference, this chapter generally will refer to architects only.

2. Owner's Objectives

With respect to the architect, the owner's basic expectations are that the project will be designed in accordance with the owner's aesthetic and financial objectives. A less well-known, but related and critical expectation of the owner is that the architect will participate in the administration of the construction contract to assure that the design objectives of the owner are fulfilled. It is this contract administration role that architects typically downplay as a component of their services, while, from the owner's standpoint, it may be one of the most important services, if not the most important service, the architect provides, especially in the case of owners inexperienced in construction projects.

Many owners are engaged in a onetime or sometime event that is far removed from their primary areas of focus. They do not have the resources to have design or construction expertise in-house. In that case, owners often view the architect as their representatives and protectors through the construction process. Unfortunately, most form contracts rarely require the architect to provide that level of support. Accordingly, the first step an owner should take is to assess its own needs and to determine how much assistance it will require from its architect.

The obligations and requirements of owners may vary greatly depending upon the nature of the owner—whether it is public or private or whether it is publicly traded or a small business. Some of the distinctions flowing from the type of owner will be addressed more in Section IV of this chapter. Before addressing those distinctions, however, this section, and the section that follows concerning contractors, largely will focus on areas of common concern.

B. Program

The owner, generally, is charged with communicating its needs to the design professional. Many standard contracts impose that duty on the owner. If the owner is relying on its architect to help it determine what the owner should or should not include in its program, then the agreement should explicitly state this fact. It is not uncommon for an owner to think it is receiving this service when the contract does not clearly require that of the architect.

C. Scope of Services and Related Issues

The scope of the services that the architect provides is one of the most important and least understood areas in the construction process. Owners are so anxious to begin the design process, that they often execute form agreements without first consulting their lawyers. This is dangerous because many form agreements go to great lengths to define as narrowly as possible the basic services of the architect, while making demands on the owner that many owners simply are not prepared or able to meet. One of the most important challenges to the practitioner is to ensure that the architect's scope of services is broadly enough defined so that it includes all of the services necessary to complete the project. Otherwise, the owner will be required to pay for additional services not covered by the architect's basic fee. A few of the key issues that every owner should consider when entering into an agreement with a design professional are identified below.

1. Designing to a Budget

If an owner has a budget that it cannot exceed, it is important to require the architect to design a project that can be built within the fixed limit of construction costs. If the bids or cost estimates by the contractor exceed the budget, the architect should be required to redesign the project without additional compensation.

2. Compliance with Laws

Many owners are amazed to discover that most form contracts do not require the design professional to comply with laws in the delivery of its services. Even more surprising, many contracts do not require the plans and specifications to comply with laws and codes, such as the Americans with Disabilities Act or local building codes. Allocating responsibility for the compliance with laws is particularly important, because the law is well settled that unless properly disclaimed, the owner warrants the sufficiency and adequacy of the plans and specifications to those that rely on them. The case that is most prominently cited for that proposition is *United States v. Spearin*, 248 U.S. 132 (1918). The holding in that case is generally referred to in construction practice as the Spearin doctrine.

3. Obligations of the Owner

The owner should carefully review and limit the obligations imposed upon it to those it is capable of undertaking. In many form contracts, if these provisions are not modified, the owner could unwittingly compromise its ability to enforce the contract against the architect. For example, AIA Document B141-1997, *Standard Form of Agreement between Owner and Architect* ("B141"), requires the owner to provide "full information in a timely manner regarding requirements for and limits on the project."[1] The architect, not the owner, may be better suited to understand what is required for the project. The burden should not be shifted to an unprepared or ill-prepared owner.

Another provision of the B141 requires owners to furnish the services of consultants not described in the contract. If all necessary consultants are not designated as part of the architect's basic services, the owner could be required to pay for these consultants or any additional services outside the architect's basic fee.

---

1. The American Institute of Architects, AIA Document B-141-1997, *Standard Form of Agreement Between Owner and Architect* (1997) at Article 1.2.2.1.

Owners often are required to review and approve items within a specific period of time or "so as not to delay" the project. The owner must make sure that it has both the expertise and the staff to fulfill those obligations. In addition, it may be inappropriate to require an owner to approve a particular item, and thereby to assume liability for the approved item's outcome.

There are certain obligations that are appropriately placed on an owner. The owner should provide the architect with legal information concerning the ownership of the project and the use of the site. The owner's budget should be realistic, and the owner and design professional should work together to reach that result. The owner should commit to pay for services that are properly performed within an agreed upon time period. If the owner cannot pay the design professional until it receives financing, and services will be needed before that time, the agreement should define the payment arrangements clearly and fully.

4. Construction and Post-construction Services

There are many services that the owner, and the owner's lender, may expect the architect to perform during the construction of the project. Often the owner is expecting more from the architect than the architect is committing to provide. It is also critical to understand the obligations of the other parties to the construction process, such as the contractor, to make sure that there are no gaps in providing essential or required services.

One area that often results in the greatest difficulties is the monitoring and supervision of the construction. The standard form agreements often provide that the architect will visit the site only periodically and only to observe the project in the most general way. The architect usually will disclaim any responsibility to evaluate the quality of the work and for site safety. These obligations are assigned to the contractor. While the architect may not be willing to undertake responsibility for safety, it may—for a fee—be willing to increase its role in monitoring and inspecting the work. This action may be important if the owner has limited staff or capability for performing these functions. The owner and the architect must determine how frequently site visits should occur and for what the architect is responsible on those visits.

The owner also should determine if it wants the architect to review the contractor's payment requests and to certify their completeness and accuracy. If the owner requires this service, the architect should be obligated to make its recommendations to the owner within a specific period of time. Similar concerns arise if the architect will be required to review the contractor's requests for change orders or for information during the project. The owner also should consider making the architect's certification simply one step in the payment process for the contractor, so that if the owner disagrees with the architect's certification, it is not obligated to pay the contractor.

Because the project as it is finally built rarely is identical to the plans prepared before the project is built, it is important for owners to obtain a set of plans reflecting the final project. These plans are called "as-built" drawings. They may be annotated plans created as construction progresses or new plans created from the annotated set. No matter what the form, the responsibility for preparing and providing the as-built plans should be clearly designated.

The architect also may be involved in assembling equipment and project operations manuals. The owner should make sure that those services are included within the architect's basic compensation, and that the manuals are received before final payment is made.

The architect's services should not always stop when a project is, or appears to be, completed. It may be appropriate to require the architect to return for one or more post-completion walk-throughs, particularly in advance of when any applicable warranties or obligations to correct defective work may expire.

5. Architect's Authority

The architect's authority to bind the owner throughout the project should be carefully analyzed. The owner should consider whether it wants the architect to have the authority to reject work, to receive and filter all communications between the owner and the contractor, and to judge the compliance of the owner and the contractor with their contract obligations. The level of the architect's involvement will depend upon the ability of the owner to assume those responsibilities and the trust and relationship between the owner and the architect.

D. Standard of Care

Another key issue is the standard of care required of the architect. If the standard of care is not described in the agreement, the law of the applicable jurisdiction will determine the standard of care. Generally, the law will define the standard as that of the average professional in the same area. If the architect was engaged because of an expressed proficiency, or if the project is very complicated, that standard may not be sufficient. A higher degree of care may be more appropriate; if so, it is critical that the architect's professional liability insurance will cover claims based on the increased standard of care.

E. Ownership of Documents

Another very critical area for owners is the ownership of the design documents. Almost all of the form documents provide that the architect retains all incidents of ownership of all documents, including copyrights. This can be particularly problematic if the architect is terminated or leaves the project before it is completed. On the most basic level, owners often feel that because they are paying for the plans, they should own them. This is contrasted with the legitimate concern of the architect that the work be paid for and that the plans not be used improperly, thereby exposing the architect to increased liability.

Both the owner and the architect may be concerned about the reuse of proprietary information or the reproduction of something that is intended to be a feature unique to that particular project. The provisions relating to the ownership and use of the plans should be carefully reviewed to accommodate the legitimate concerns of all the parties. The owner's lender also may want a role in this process to ensure that it has access to the plans to complete the project if the owner defaults in its obligations.

Work that may be copyrighted is no longer just in the form of paper drawings. Electronic media raises new issues of what is a completed work, and who owns it, especially where more than one party may make changes to the work. In addition, finished buildings may be subject to copyright, and art work incorporated into buildings may also be subject to protection. All parties should be aware that copyright and other applicable areas of the law are informed both by regulatory schemes and the common law. Specific language may be necessary to accomplish the parties' intended results.

A related and newly emerging area of the law concerns the rights and obligations of the parties using project extranets. The ability to keep information confidential, the responsibility for losses if the system fails or the independent provider goes out of business, the potential for corruption of information when documents are transmitted electronically, and the protection of intellectual property are just a few of the issues that the practitioner must address when project extranets are used.

F. Licensing

It is important to require the design professional to be properly licensed. Most jurisdictions have comprehensive statutory requirements mandating licenses in order to perform design work. Some jurisdictions limit what services engineers, as opposed to architects, may provide. All prohibit the practice of architecture without a license. This issue often arises where portions of the design are delegated to parties other than the licensed architect.

Because construction projects are becoming increasingly complex, many components of a project are now designed by contractors or their specialty subcontractors or material suppliers, rather than by the architect. The architect will establish criteria or performance results that must be achieved, but the method of achieving those results will be delegated to other parties.

This allocation of responsibilities is not necessarily inappropriate, but it does raise certain issues. For example, the owner should confirm that the party providing the services is licensed to do so, and is properly insured. The owner may want a direct right of access against the third party. The owner also needs to determine who is monitoring the work and who ultimately is responsible for the design.

G. Insurance

Because the design process is far from infallible, insurance is one of the most critical elements of a construction project. Many form documents do not require the architect to carry insurance of any kind. While limits may be subject to availability and vary depending upon the size and complexity of a project, architects errors and omissions (E&O) insurance always should be required. Because E&O insurance is typically issued on a claims made rather than an occurrence basis, a "tail" or extended reporting period may be needed to cover claims that are discovered after the project is completed. In addition, architects' E&O policies typically renew on an annual basis, which could result in only a one year window of opportunity for claims to be made by the owner before they are barred by a policy concept known as the "retroactive date."

The architect's policies also may cover a myriad of projects and coverage may be exhausted before an owner can make a claim with respect to its project. Owners may want to require architects to maintain project specific coverage to avoid this problem. Owners also may want to require architects to maintain coverages such as "valuable papers," so that designs may be reconstructed if they are lost due to an accident or casualty, such as a fire or power failure.

H. Indemnification

A well-drafted indemnification provision is critical. The indemnification should be sufficiently broad to cover errors and omissions and other negligent acts of the architect, as well as a breach by the architect of its obligations under the contract. It is important that the indemnification obligations be covered by the architect's commercial general liability insurance.

I. Limitations of Liability

Recently, much has been made of issues regarding the limitation of liability of the parties to a construction project. There is a trend in many form contracts requiring both parties to waive consequential damages. While the definition of consequential damages varies from jurisdiction to jurisdiction, consequential damages are generally defined as those that flow from a wrongful act, but which are not the direct and immediate result of it. For example, if the architect delays the project by an inexcusable failure to produce plans in a timely manner, direct damages might be the increased amount of interest that an owner is required to pay to its lender because of the delay. An example of an owner's consequential damage may be the loss of expected tax credits for failure to place the project in service by a legally required date.

A mutual waiver of consequential damages is not always appropriate. The owner should evaluate which party is in the best position to control the risks and may not want to excuse that party from the consequences of its failure to do so. Unlike those facing the owner, the consequential damages that the architect faces may be dealt with more easily by contract provisions that protect the architect. For example, the contract could require that a certain percentage of lost profits be paid to the architect if the architect is terminated early without cause. The contract also can provide for payments if the architect is delayed without cause beyond a specified time period. In these cases, it may be appropriate for the architect to waive its rights to consequential damages against the owner without the owner waiving its rights against the architect.

J. Dispute Resolution

Many form contracts provide for some kinds of alternative dispute resolution. Without making a judgment as to whether a particular form of alternative dispute resolution is appropriate, it is important for owners to review both the type of process that is required and the procedures governing the process. The owner may want to retain the right to determine which claims are mediated or arbitrated and which are litigated.

One area of concern is whether an owner must engage in repeated proceedings, when similar, prior proceedings have resulted in little benefit. Also, it is important that all the potentially relevant parties to the arbitration or mediation are allowed or even required to be a part of the process. While that might sound like a simple truism, the B141 prohibits consolidation and joinder. Accordingly, an owner may be engaged in disputes in more than one forum, with resulting additional costs and the possibility of inconsistent results. Finally, the owner should assess whether the arbitration decision should be final and binding or may be subject to appeal.

Lenders often are unwilling to participate in arbitration and other forms of alternative dispute resolution. Owners may want to provide that their lenders are not bound by any of these provisions.

K. Termination of the Architect's Agreement

The owner should look at both the reasons the architect may terminate the agreement or suspend its work, and the reasons the owner may terminate or suspend it. It is important for the owner to be able to terminate for both cause and for convenience. The term "for convenience" may be a misnomer. It generally does not mean the owner is terminating on a whim, but rather, that the owner cannot fund the project, or that other problems make it difficult for the owner to continue with the project. It is also an important tool if the architect's and the owner's relationship has broken down, but the breakdown is not sufficient to justify a termination for cause. This type of termination is more in the nature of a no-fault divorce.

In the event of a termination for convenience, it is clear that the architect should be paid for its work to date, and should receive something that may be characterized as "termination expenses." These "termination expenses" could be something additional to compensate the architect for the loss of some of its expected profits; it should not, however, be all of the profit that the architect claims it otherwise would have made, because the architect is being relieved of any risk from that point forward and is free to engage in other jobs.

No matter what type of termination is involved, the architect should be required to cooperate with the transition to the new design professional, if there is one. Provisions also should be made for the continued use of the architect's documents. Where there is a termination for cause, the architect should be liable for all damages incurred by the owner, and should not receive any further payments until the owner can quantify those damages fully.

Sometimes a project is not terminated, but must be suspended because permits take longer to obtain than initially contemplated or regulatory or other unexpected events intervene. Suspensions should be contemplated in the agreement, and the costs of delay, if any, allocated clearly. The owner should be able to require the architect to resume work if the project recommences within a stated period of time.

L. Release of Lien

Many states provide a lien right to architects that is similar to a mechanic's lien, where the architect is providing any construction administration services. This right is often overlooked by owners and their counsel. In states where architects have a right to lien, prudent owners should require the architect to sign a partial release of lien at the time of each payment of the architect's fee, in generally the same form that is required of contractors. The architect should also deliver a release of its lien rights at the end of its services. In a minority of jurisdictions, the owner may be able to require the architect to waive its lien rights in advance of the performance of the work. The practitioner should become fully familiar with the lien laws in the relevant jurisdiction.

III. RELATIONSHIP WITH THE CONTRACTOR

A. Introduction

The owner's expectations of the contractor in the most basic sense are that the contractor will construct the project in a way that satisfies the owner's objectives for scope, time and cost. Most owners also expect that the contractor's commitments are based on a complete evaluation and understanding of the physical conditions under which the project will be constructed and a thorough analysis and understanding of the plans for design of the project. Moreover, most owners generally expect that in constructing the project, the contractor will look out for the owner's well-being by, for example, calling the owner's attention to design defects noted by the contractor. While these less precise expectations are natural ones, they are dangerous ones because they are generally not obligations of the contractor unless expressly specified. In fact, construction contracts are one of the clearest examples of a relationship where an owner's expectations—even very natural ones—must be expressly spelled out in order to stand a chance of being fulfilled.

B. Types of Contracts

Construction contracts are often described by the method chosen for payment of the contractor. The most common varieties include lump sum, cost-plus-a-fee, guaranteed maximum price and unit price. Each is discussed below.

1. Lump Sum

A lump sum or stipulated sum contract is one in which the contractor agrees to build the project, including supplying all labor and materials, whether by the contractor or by subcontractors, for a specified dollar amount (the "lump sum"). The owner is obligated to pay the contractor the fixed amount no matter what it costs to finish the work. The contractor's fees, profit and overhead are not separately stated.

The owner pays the contractor based on the amount of work that is completed, and does not pay attention to the actual costs incurred by the contractor. Thus, if the contractor is very efficient, it can reduce its costs and make a larger profit. Because the contractor has committed to a specific cost, and the owner has agreed to the costs regardless of what the project actually costs to build, the owner is not entitled to review the costs that the contractor incurs during the project, except, perhaps, when change orders are requested.

This approach is often used in government contracts and where the contract is put out for competitive bidding. With this approach, the plans and specifications generally are assumed to be complete or close to complete before a price is given by the bidding contractors.

2. Cost Plus

The second approach is the cost of the work plus a fee. Under this scenario, the contractor is reimbursed for its actual costs (e.g., costs of labor, materials, equipment and insurance), and it also receives a fee. The fee is usually some percentage of the cost of the work. This approach requires that the owner review the costs the contractor has incurred, and the construction contract must carefully define what costs are reimbursable.

There often is a great deal of negotiation as to what is reimbursable and what is not. The contractor will want to recover as much of its home office costs as possible. The owner should make sure that these costs are considered part of the contractor's general overhead and thus included in the fee. Many other questions will arise. For example, if a subcontractor makes a mistake, should the cost of correction be reimbursable? Should there be limitations on reimbursement for rental equipment, especially if rented from an affiliate of the contractor? What about computer costs? Should they be included in the fee? Whether included as part of the contractor's fee, or separately stated as a cost of the project, the contractor will also typically be paid for "overhead" and "general conditions". A share of the contractor's general overhead costs may be fairly apportioned to the project in question, as well as reimbursement for the typical construction tasks and materials performed or supplied by a contractor that are essential to the smooth running and successful completion of the project, but which generally defy precise categorical description. The latter items, which may include such things as temporary utilities, barricades, hoists and supervision by the contractor's employees, are called "general conditions items".

Because under this method there is no cap on costs, it is the least likely to be used. It is used most often when the plans and specifications are incomplete and therefore a lump sum or guaranteed maximum price cannot be established. The cost of the work approach is more difficult for an owner to administer than the lump sum approach, because the owner must examine each payment requisition in great detail to make sure that only permitted costs are being reimbursed. With the lump sum method, the owner only needs to pay attention to how much of the work has been completed. In this type of contract, the owner should be permitted to see all of the contractor's charges and expenses.

3. Guaranteed Maximum Price

In a pure cost-plus-a-fee contract, at the outset there is theoretically no cap set on the ultimate project price because the project price will be the result of all the individual subcontract costs, then unknown, plus the contractor's fee. In a guaranteed maximum price (GMP) contract, the contractor will agree with the owner that the costs of the work, plus the contractor's fee, will not exceed a guaranteed maximum dollar amount. In other words, while the sum of the actual costs plus the contractor's fee may result in a project price less than that guaranteed maximum amount, the sum total of those costs plus the contractor's fee cannot exceed the maximum price guaranteed by the contractor. Unlike the lump sum approach, where the owner's cost is known at the time of contracting, and unlike the pure cost-plus-a-fee approach, where the owner's ultimate cost for the project is unknown at the time of contracting, under a GMP contract, the theory is that the owner will know at the time of contracting that it may pay less than, but never more than, the GMP.

Often, as an incentive to the contractor, if the final cost of the work and fee are less than the GMP, the owner will share some of the savings with the contractor. The owner also usually will agree that the contractor's fee will apply to change order work. As with the cost-plus-a-fee contract that is not subject to a cap, the contractor's books should be open for review by the owner.

One issue that often arises with a GMP contract is whether the contractor is guaranteeing just the overall price, or whether it is, in addition, guaranteeing the various components of the price (e.g., that the excavation, mechanical systems and plumbing systems will each not exceed a certain price). Most contractors will refuse to guarantee line items of price. They will want to be able to transfer savings from one line item to areas where there have been overruns. Most owners will agree to this concept, at least within a certain range, with one exception. Owners should impose a cap on the general conditions costs, because those costs are a potential additional profit center for the contractor.

Many owners feel that GMP contracts give them the best chance to achieve the lowest project costs, because they know what the contractor's fee will be and by requiring the contractor to obtain competitive bids for the work of the trades, have some control over other costs. GMP contracts are usually (although not exclusively) found on large, sophisticated projects in the private sector. Most owners negotiate with a preselected contractor the terms of a GMP contract, rather than bidding out the contractor's portion of the job. A GMP contract can still be used where the plans and specifications are not far enough along for a meaningful GMP to be established. In such a case, the contract can provide for the work to begin initially on a pure cost-plus basis, or with an artificially high GMP, with a more accurate GMP to be established later, when the plans are closer to completion. Of course, that method of contracting leaves a risk that the parties will not be able to agree on the more accurate GMP. This issue highlights the fact that GMPs are often given with respect to plans and specifications that are incomplete. Although a contract may have very strong language to the effect that no increase in the GMP will be granted to the contractor if the final plans are consistent with the initial less developed plans, it is often very difficult to defend against a claim that the final plans have increased the scope of the job and hence the contractor is entitled to more money. The GMP is only as good as the plans on which it is based, and owners fool themselves if they think a GMP is meaningful if it is based on inadequate plans.

4. Unit Prices

Although it is rare that an entire contract will be based on unit prices, this approach is seen in construction projects directed at the construction or installation of one or a few specified materials, such as laying pipelines or the construction of sidewalks. In a unit price contract, the contractor agrees to specified unit prices for specific items of material and labor, and the ultimate total contract price becomes the sum total of the number of "units" of that material and labor ultimately required to build the project.

5. Allowances

Many contracts also will provide "allowances" as part of the contract price. Allowances are an amount of money that the contractor has allocated to a particular item or portion of the work for which it does not have a detailed bid or estimate. In other words, the owner is "allowed" up to the amount specified to require that item or work from the contractor. If the actual cost of providing or performing that work exceeds the allowance, the lump sum or GMP will be increased by the overage. It is critical for an owner to make sure that an allowance number is a realistic estimate of the likely cost of the item and that it includes all of the work, such as labor, materials, taxes, the contractor's fee or profit and overhead.

C. Contract Documents

The term "contract documents" means something very different in the construction industry from its meaning in other areas of contract law. It is the term used to identify the materials, in whatever form or medium, that describe both the work that is to be performed by the contractor and the relationship between the owner and the contractor. If the parties are using AIA form documents, at the start of construction, the term "contract documents" might include the following:

1. an agreement between the owner and the contractor containing the essential elements of the owner/contractor relationship, such as the identity of the parties, a description of the work, the price, the time for completion and a list of the contract documents;
2. general conditions of the contract for construction, which expands upon the information provided in the agreement between the owner and the contractor and provides more complete contract terms governing the parties, such as the owner's obligations to the contractor, the services that the contractor will provide, insurance and bonding requirements, changes in the work, payment procedures, and although the architect is not a party to this document, the construction administration services that the architect will provide;
3. supplementary conditions of the contract, modifying or supplementing the agreement and general conditions of the contract;
4. drawings prepared by the design professional, which are the graphic and pictorial expressions of the work to be performed, and which may be in electronic as well as printed media;
5. specifications, which are the written technical requirements for the materials, equipment, systems, and standards for the work; and

6. addenda, modifying or supplementing any of the above issued before the contract is executed.

After the contract is executed, other items may be added to the list of contract documents. For example, modifications to the contract or change orders issued concerning the scope of the work may be issued and become a part of the contract documents.

It is essential that the owner understand what items constitute "contract documents," both in the beginning of the project and as the project progresses. Because the documents may be produced at different times and by different entities, it is almost impossible for them to be free of inconsistencies or contradictions. Therefore, one critical issue is the order of hierarchy between and among the various items constituting the contract documents. An example of one opportunity for inconsistent provisions can be found in the first division of the specifications. The specifications almost always contain provisions purporting to establish general requirements relating to the use of the construction site, the processing of change orders, responsibility for site safety and other procedures. These topics also are set forth in the general conditions to the contract for construction. The specifications, which often are produced by the architect and rarely reviewed by the owner or the owner's lawyer, may contain provisions that contradict provisions that were heavily negotiated between the parties.

As previously discussed, the contract documents also often bind the owner to provide certain services of the architect and other consultants. The owner must be certain that it has agreements to provide those services from the other individuals or entities who are not parties to the contract documents.

Owners also should avoid incorporating items into the definition of contract documents that may modify the obligations imposed on the contractor. For example, contractors may ask that their bid proposals and shop drawings be included in the definition of contract documents. Bid proposals often contain qualifications and limitations on the contractor's responsibilities to which an owner has not agreed. Shop drawings, which will be discussed later in this chapter, may modify the plans approved by the owner in ways which the owner may not intend.

The owner should require the contractor to review carefully the contract documents and identify inconsistencies or areas that are unclear to the contractor. Contractors often resist this obligation, claiming that it is the architect's sole responsibility. A good contractor should be willing to engage in this exercise if the scope of its review is limited to identifying those items that a reasonable contractor, experienced in the type of project that is being undertaken, would discover.

D. Scope of the Work

In the simplest terms, the contractor is charged with completing a project that is described in the plans and specifications and the other contract documents. It is critical that the work include all labor, materials, equipment, taxes and any other items that are necessary to produce the completed project. If the definition is not sufficiently comprehensive, the contractor will request or demand an increase in the amount of its payment.

Owners also should understand that plans and specifications, by definition, will omit some details of the project. The contractor will suggest that any omission should entitle it to more money. The architect will argue that the contractor should have known what was intended from the documents, even if the architect omits some detail. The owner can be caught in the middle, unless the contract allocates responsibility in these circumstances. One of the most common ways to deal with this issue is to require the contractor to perform work that reasonably is inferable from or intended by the contract documents.

E. Time

In addition to issues regarding the cost of the work, and the scope of the work, another major issue for an owner is the time for completion of the work. Unlike many other areas of the law, the construction industry deals with two different types of completion. The first, called "substantial completion," describes a time and a state of completion that is less than fully finished. How much less than complete often is the subject of controversy. Items that are not completed at the time a project is considered substantially complete are often referred to as "punchlist" items. The definition of substantial completion is very important because it is often the time when any remaining retainage must be paid (the topic of retainage will be addressed in the section on progress payments), when statutes of limitation begin running (at least for defects that are actually or constructively known) and when warranties commence.

In most construction contracts where the plans and specifications are reasonably complete, the contractor will agree to reach substantial completion of the entire project by a certain date. The owner should make sure that the definition of substantial completion includes all of the elements necessary for the owner to use and receive the full benefit of the project. For example, the definition of substantial completion in most form construction contracts is very general. If the owner, in order for rent to commence under a lease, needs to have the tenant sign off on the work and receive a certificate of occupancy, or to have landscaping completed, the owner should include those items as part of the definition of substantial completion.

In some circumstances, interim milestone dates may be important. Major tenants often require their landlords to achieve completion of portions of work by specified dates. If the landlord fails to meet any interim date, the tenant can terminate its lease obligations. In that event, the landlord/owner must pass the obligation to meet interim dates on to the contractor.

Another issue that almost always arises when discussing time or schedule is the definition of "force majeure", a term which means causes beyond the owner's or contractor's control. Most construction contracts extend the required substantial completion date for delays beyond the contractor's control. The owner can expect to pay a hefty premium in order to get a contractor to take the risk of all or most of the force majeure events, so it is rare. Thus, the owner will want a very tight definition of force majeure, while the contractor will argue for a very broad one. For example, the contractor will want delays due to inclement weather to be defined as force majeure event. The owner likely will want the contractor to demonstrate that the weather substantially differed from what normally would be expected in the locality of the work at that time, based on historic weather data. The contractor may want labor disputes to constitute force majeure, while the owner may consider that to be something within the contractor's control. Sometimes they compromise: an industry-wide labor dispute will be considered force majeure while a dispute limited to the contractor's workforce will not.

Another issue related to time is what happens if the contractor is delayed in the work by the owner or another party. Delays cost contractors money because they have costs (*e.g.*, equipment rental) that they may incur whether or not they are working. Obviously, if the contractor has to stop work for a period of time and then recommence so that the length of the job increases, the contractor will incur increased costs. Many owners try to get contractors to agree that if they are delayed by the owner or the architect, their sole remedy will be an extension of time. This is called "no damages for delay." A carefully drafted "no damages for delay" clause is enforceable in most states. Contractors do not like to take this risk. If they do, they will charge a premium for it in their price. If an owner agrees to pay for some costs of the delay, it should be wary of amounts claimed for the contractor's alleged increased overhead costs and any claimed loss of opportunity.

In many contracts, particularly ones with public entities, damages are imposed for the failure of the contractor to substantially complete the project by a certain date. The damages imposed often are referred to as "liquidated damages." It is important to carefully describe the circumstances for which liquidated damages can be imposed.

Liquidated damages clauses generally will be upheld only if at the time the contract was made, it appeared that the precise amount of harm flowing from the breach would be difficult to estimate. This type of agreed damage clause is extremely helpful where damages from the breach may involve intangible, remote or hard to prove elements, but where such damages nevertheless are real. Liquidated damages provisions will not be enforced if they are deemed penalties, such as where the agreed upon amount is disproportionate to the amount of damages that the parties reasonably contemplated might ensue from the breach at the time the contract was entered into.

On projects where it is particularly critical to the owner that the project is completed as quickly as possible, the owner may include financial incentives for early completion. The incentives are most often included in the contract as "bonus" payments where the contractor is entitled to additional money if the project is completed and ready for occupancy prior to the contract completion dates. In including such financial incentives, it is important to carefully define the condition that is deemed to be completion for the purpose of determining when the contractor will be entitled to the additional payment.

F. Progress Payments

It is surprising how often owners neglect to conform the payment provisions of the contract with the lender's requirements or the owner's ability to review requests and obtain necessary funds. The owner must determine whether it needs outside help, such as help from the architect or a project representative, to review the sometimes technical and complex payment requisitions. The owner also should be aware that in many jurisdictions there are statutory requirements imposing the time periods within which payments must be made or interest and penalties can accrue.

1. Draw Requests

Progress payments are typically initiated by a request of the contractor, called "an application for payment" or "draw request." Whatever the nomenclature, from the owner's standpoint a draw request should contain at least the following items:

a. the amount requested;
b. a substantiation of the amount due based on the method agreed for determining progress payments;
c. the architect's approval of the draw request, or recommendation to the owner; and
d. a list of the subcontractors who have provided work and materials during the period for which the progress payment is to be made and mechanic's lien waivers or releases from those subcontractors, and the contractor itself, for the work and materials represented by the progress payment in question.

In other words, the owner should be assured that it is paying the correct amount and that, upon payment, the work paid for by the progress payment will be fully paid for, without lien or claim by anyone involved.

In some cases, progress payments are based on a "schedule of values," or a schedule showing how much various portions of the work are worth in terms of payment. Progress payments can be made, then, on the basis of the amounts allocated to the various items of work as they are completed, or according to the percentage completed on any otherwise regularly specified payment date. A second approach is to make progress payments based on the cost of the work and materials to the date of payment, plus an allocable portion of the contractor's fee in a cost plus contract.

2. Retainage

Another unique aspect of the payment provisions in a construction contract is the concept of retainage. Retainage is an amount withheld from each of the payments due to the contractor, which is not paid until a specified time, often at the end of the construction. Retainage is viewed as an added incentive to complete the work on a timely basis. In addition, that same fund of money can be used to satisfy unexpected mechanic's liens or claims which may arise, or may be applied against any default by the contractor during the job.

The tension between the owner and the contractor regarding retainage is easy to understand. Owners typically attempt to retain between 5% and 10% of each progress payment. Contractors resist the notion of any retainage or attempt to negotiate the retainage figure to be as low as possible. A common request by contractors with respect to retainage is to retain the agreed upon percentage from each progress payment until the job is a certain percentage complete—normally 50%—and then to eliminate the retainage feature, provided certain preconditions are met. The idea behind this approach is that the owner is at progressively less risk as the job is completed and that, after a certain point, the overall amount of retainage can be reduced given satisfactory progress of the job. In other words, if 10% of a progress payment were being withheld until the job was 50% complete, and then retainage was reduced to nothing, the owner would, in effect, end up with a retainage of 5% of the job overall.

3. Contractor's Right to Stop the Work

A final basic point involving progress payments concerns the issue of the contractor's rights in the event the owner disputes the amount the contractor has requested. Contractors frequently request that they be able to stop work if the owner does not pay the amount requested, whether or not the owner has disputed that amount in good faith. Owners resist this request because it gives the contractor tremendous leverage over the owner in that the contractor may be able to delay the job—and subject the owner to attendant delay costs—while a dispute is resolved. Any stoppage in a job may be a default under the owner's loan documents. Owners tend to prefer an approach, therefore, under which the contractor is paid that portion of a progress payment request which is not disputed in good faith by the owner, and has no right to stop work for nonpayment of the remainder while the dispute is being resolved.

G. Shop Drawings and Project Design

Shop drawings are illustrations prepared by a contractor or subcontractor of a specific aspect of work. For example, the plans and specifications may describe a particular type of air conditioning system that is to be used, but may not show how the ductwork for the system will fit within the ceiling. A shop drawing prepared by the HVAC subcontractor will set forth the details of the installation. The shop drawing is then submitted to the architect for the architect's review.

It is a major point of controversy in the construction industry as to what the scope of the architect's review of shop drawings should be. The contractor and owner want the architect to "approve" the shop drawing, thereby assuming at least some of the responsibility for its effectiveness. The owner also will want the contractor to retain liability jointly with the architect. The architect wants to review the shop drawing only for limited purposes, such as to ensure that the proposed detail generally conforms to the architect's intended design, and will not want to issue an approval. However the issue is resolved, the contract should clearly define the respective roles of the parties. If the contractor is the responsible party, the owner should make sure that the contractor's insurance covers errors and omissions in the design and that the contractor is licensed to provide any design work that may be involved.

H. Changes

Changes in the work may occur in a number of different ways. In many instances, a change in the agreed upon work can only be made by a contract amendment which, in construction terminology, is normally known as a "change order". Change orders typically describe agreed upon changes in the work and the cost and schedule implications of those changes. Since execution of a change order by both parties presupposes agreement to the change in the work and the cost, schedule or other implications of the same, there should be few, if any, issues to contend with when the system works as contemplated.

Owners often need to be able to order a change even where the parties have not agreed to all of the terms. In the AIA documents, that type of change is called a "Construction Change Directive." If construction change directives are provided for, the contract should set forth the method of how an agreement ultimately will be reached.

Another issue is whether the architect has independent authority to direct "minor" changes in the work, without the owner's or contractor's consent. Minor changes should not change the cost or the schedule. It goes without saying that "minor" is a subjective judgment, and further that the cumulative effect of a number of minor changes can be major. Accordingly, most owners resist giving the architect this unfettered authority.

I. Safety of Persons and the Work

A contractor is generally responsible to insure the safety of persons on the job site and of the work. Part of this responsibility is subsumed within the contractor accepting the requirement to determine the "means, methods, techniques, sequences and procedures" of the work. It is important that the owner does not agree to perform supervisory services over the project, to approve the contractor's safety plans or to take on other activities that may make the owner jointly responsible for safety with the contractor.

J. Dispute Resolution

Many forms of alternative dispute resolution are available and are finding favor in the construction industry. These include arbitration, mediation and partnering. Partnering involves all of the parties to the construction project agreeing to be involved in a process that starts before the work begins. The process generally is led by a neutral party, and the goal is to avoid problems and to intervene early to minimize the number and severity of the problems.

The issues discussed in Section II.J. concerning arbitration or disputes with the architect apply to disputes with the contractor as well. Owners must be particularly cautious when the AIA form documents are used. In addition to prohibiting consolidation and joinder of the parties, the AIA forms give the architect a relatively central role in making initial determinations as to most disputes and controlling determination concerning disputes with respect to aesthetic effect. Most owners modify or eliminate these provisions.

K. Indemnification, Insurance and Bonding Provisions

Entire volumes are written about insurance and bonding in the construction industry and the law surrounding indemnification and liability provisions. A paragraph cannot do the subject justice. A broader discussion of insurance issues is included in Chapter 12. The most important thing for a beginning practitioner to understand, however, is that these provisions are critical and must not be treated as "boiler plate."

1. Insurance

The contractor should be fully and thoroughly insured for its own liability and as to its own persons and property. In addition, someone—either the owner or the contractor—must procure and maintain in effect "builders risk" insurance. This basic form of insurance covers damage to or loss of the job in progress. Without it, if a job in progress is fully or partially destroyed, the owner may simply be out of luck.

Do not get caught in the trap of confusing the obligation to pay for builders risk insurance with the obligation for procuring it and maintaining it in effect. It will almost always be paid for by the owner, either directly or indirectly, as a portion of the construction contract price. But someone must get it and be responsible to maintain it.

The owner should also pay attention to the amount of any deductibles and who has responsibility for bearing any loss not covered by the deductible. Insurance must run to the benefit of any lender and the lender may have specific requirements as well.

On almost any insurance issue, it is worth consulting with insurance professionals or professional risk managers to verify that the requirements specified by the contract are adequate to protect the owners' interests and will work as the parties expect them to. Construction insurance is a specialized area of the insurance industry and the owner's risk manager may not be qualified to determine what coverage is adequate.

2. Bonds

In many projects, particularly public ones, performance and payment bonds are required to guarantee the contractor's performance. In very general terms, a performance and payment bond provided by a surety is designed to assure the performance of the contractor in carrying out the work and in the payment of its subcontractors. Greater coverage of the details of a surety's role are addressed in Chapter 5 of this book.

In design-build or fast-track projects, where designs are not completed before construction begins, bonds for the full amount of the project may not be available. In that case, incremental bonding may be the only alternative, requiring careful monitoring by the owner. If the project becomes troubled, the surety may refuse to provide further bonding increases. In addition, the bond may not cover the design obligations of the work.

3. Indemnification Provisions

It is critical that the contractor indemnify the owner for any damages caused by the contractor or anyone for whom the contractor may be liable, and for losses arising out of a breach of the contract. If an owner desires to be indemnified even where the owner was partly at fault, the language of the indemnification must be clear and unambiguous to be enforced.

L. Environmental Issues

Whether a project involves new construction or is a renovation of existing facilities, the risk of encountering environmental hazards is always present. The owner must assess the likely risks and the entity in the best position to control or insure against those risks. The contractor should be required to notify the owner immediately if it encounters any hazardous conditions and should stop work until the owner can determine, with its consultants, the best course of action. Unless mandated by law, the contractor should not notify public officials until the owner has had an opportunity to review the circumstances and decide whether notice is necessary and, if so, who should undertake that responsibility.

If environmental remediation is part of the project, the owner must make sure that the contractor is properly licensed and insured and that the insurance covers the contractual indemnification of the owner by the contractor. The owner may also want to consider purchasing its own insurance to cover environmental risks on appropriate projects. Many such products are now available.

M. Warranties and Correction of Defective Work

Many contracts provide express warranties, such as that the materials and equipment will be of good quality and new and that the work will be free from defects. Another common warranty is that the work will conform to the requirements of the contract documents. This warranty raises the question as to what degree of compliance or performance is required of the contractor. It is obviously in the owner's interest to have as broad and clear a set of express warranties as possible.

Construction contracts also often include an obligation of the contractor to return to the project to correct defective work within some time period (often a year) after the work is substantially completed. The requirement to correct such work should not be confused with the contractor's warranty that the work will be free from defects. These are independent covenants, and each can be enforced according to its terms and applicable law.

In addition to express warranties, implied warranties often are imported into the contract documents. The contractor may have an implied duty to seek resolution of an ambiguity which is obvious or patent, in the case of federal contracts. In residential projects, there may be implied warranties of habitability and of workmanlike construction by the contractor. It may be necessary for the owner to disclaim certain implied warranties, such as the warranty that the plans are sufficient, or the implied duty to coordinate the work of multiple prime contractors or the owner's many consultants. If not expressly disclaimed in the contract, some jurisdictions will hold that they have been made.

N. Termination

Section II.K of this chapter addresses some general issues regarding termination of an agreement between the owner and architect. The considerations discussed there are applicable to contractors as well.

Public contracts generally are required to include provisions permitting the owner to terminate the contractor without cause. The concept behind this requirement is that while construction may not literally be an *impersonum* service to the owner, an owner would like to be generally comfortable with its contractor at any given time. If the owner is uncomfortable with the contractor for whatever reason, but the contractor is nevertheless performing the literal requirements of the contract, the owner will not be entitled to terminate the contractor unless the contract provides that the owner may do so for its own convenience.

Whatever the basis for termination, to the extent that it occurs during the progress of the project, it is worth providing in detail the contractor's obligation to turn over the job to the owner or any successor contractor. Provisions addressing this matter will often concern protection of incomplete work, consolidation and protection of materials not yet incorporated into the job, cleanup responsibilities, and termination and turnover (assignments to the owner or its designee) of contracts. While some payments are appropriately made to the contractor for this type of termination, the owner must be vigilant to ensure that the contractor does not receive a windfall because of an early termination.

Where a termination is expressly permitted "for cause," the owner should carefully evaluate the reasons that are identified as just causes. For example, the AIA form documents require that the contractor "persistently or repeatedly" refuse or fail to supply enough properly skilled workers or proper materials before the owner can terminate the contract. Most owners will not want to await persistent or repeated failures, nor will their lenders.

In like manner, another basis in the AIA forms is the persistent disregard of laws and ordinances. Again, an owner should not have to suffer persistent violations of applicable laws before terminating the contract.

A general issue concerns the amount the contractor is entitled to upon termination. Approaches to this subject vary between silence (relying on common law theories of damage) to specific provisions dealing with various payment schemes when termination occurs for the owner's convenience, as a result of the contractor's default, or as a result of the owner's default. Careful thought should be given to the likely damages that may arise in each of these circumstances.

The owner also should consider whether the contractor should be required to waive any claims for consequential damages against the owner. A discussion of some of the issues to be considered in making that decision appears in Section II.I. of this chapter.

IV. WHO IS THE OWNER

There are a number of different types of owners, and their approaches to a construction project may be very different. For example, the needs of a public owner may differ from that of a private one. The requirements of an institutional owner may vary from that of a small business or residential owner. This portion of the chapter will focus on the major differences in construction projects where public or private owners are involved.

A. Public Owner

The public owner, such as a school district or municipality, will have certain legislative constraints. These may include such common requirements as the obligation to publicly bid certain types of contracts, to open bids in a public setting, to procure products from a particular locality or to refrain from procuring goods from other places, and to use only certain types of construction delivery systems.

1. Limitations on Project Delivery Systems

Many jurisdictions prohibit or place restrictions on the use of design-build contracts. One stated rationale for prohibiting design-build contracts is that placing design and construction in one contract may limit the ability to achieve the best price, which is thought to occur when as many components of a contract are bid separately as possible. Another complaint is that a design-build contract may limit accountability, because the traditional role of the architect as monitor and independent representative of the owner is compromised. Whatever the basis for objection, as design-build contracts gain more popularity in the private sector, there is more pressure for public entities to be permitted to use this form of project delivery. Each jurisdiction has its own restrictions and regulations, and it is up to the practitioner to determine what they are.

2. Employment Requirements

Public ownership often imposes limitations on employment relationships. For example, minimum or prevailing wages may be mandated. Equal employment opportunities and nondiscriminatory hiring and contracting procedures may be imposed, which almost certainly will exceed those requirements generally applicable to private entities and projects.

One of the more active areas of public debate and legal challenge is in the area of minority and disadvantaged business enterprises. This controversy is being played out in the federal courts in a series of cases known as *Adarand*.[2] The current state of the law is that federal contracts that use race

---

2. See Appendix A: Case Digest for a review of the *Adarand* cases at note 1 and text accompanying note.

as a factor in determining the successful bidder are subject to strict scrutiny. The law, however, is by no means settled. Many local jurisdictions attempt to avoid the current legal morass by making disadvantaged business enterprise requirements contractual goals rather than statutory obligations.

3. Other Regulatory Limitations

Other regulatory schemes may apply to public contracts, such as mandatory insurance, bonding and indemnification requirements. Public entities often demand that the choice of law and venue be where the public owner is located. Many public owners require, by policy, that the right to jury trials be waived. Public owners also are often mandated to require liquidated or other delay damages, and to limit the contractor's right to recover damages for the owner's delay. Public owners may be required to own any plans and generally must be able to terminate contracts without cause. Public owners may not be willing to engage in alternative dispute resolution or may have statutory arbitration or other mandated procedures. Public owners may have statutory immunities or limitations on liability, which can be waived inadvertently, if careful attention is not paid to contract language.

Public owners must be attentive to open meeting and public reporting requirements. There also may be uncodified policies or other procedures which are nonetheless applicable to public projects.

B. Private

While private owners are not faced with the same complex regulatory schemes and limitations as public ones, their needs may vary greatly depending upon their sophistication and size. If an owner has in-house staff, then it may require less assistance from its architect. If the owner only rarely, if ever, engages in construction projects, it may want to hire a project representative in addition to an architect and contractor. The important point is that the lawyers should assist the owner in assessing its needs. Among the questions the lawyer should ask its clients (and many of these questions apply to public owners, as well) are:

1. What is the nature of the owner's business? Should a separate project-owning entity be created?
2. If the owner is a publicly traded entity, is it subject to disclosure and other requirements?
3. Is the project funded from third party sources that may impose their own requirements?
4. Are there any confidentiality issues, such as new products or technologies that must be protected?
5. Is this a onetime project, or is it likely to be repeated?
6. What is the scope of the project? Has the programming been completed?
7. Has the budget been determined?
8. When does the project have to be completed? Has anyone reviewed the feasibility of the schedule?
9. Is the project owner occupied, or will it be leased or sold?
10. Who are the consultants? How are they being selected?
11. Are there market constraints applicable to the project, such as availability of materials or personnel?
12. Is there competent in-house staff with sufficient availability of time?

This list is by no means exhaustive. It is intended only to illustrate some of the issues that an owner and its lawyer should consider before a construction project is undertaken. Rarely will the lawyer have the luxury of comprehensive, advance planning with its client. It is important, however, to attempt to become as familiar as possible not only with the particular project at issue, but with the peculiar needs and interests of the client.

V. SPECIAL LENDER CONSIDERATIONS

The lender shares many concerns with the owner, but it also has a number of issues that it needs to address, on which the owner may not be focused. The lender's attorney should help the lender identify and resolve those issues. The owner's attorney assists his or her client by anticipating and dealing with the lender's concerns.

Owners and lenders often are unaware that some form documents (the AIA and AGC forms, to name two) require an owner to provide financial information to the contractor to establish that "sufficient financial arrangements have been made" to allow the completion of the project. The contractor may not only demand this information before the project starts, but may also continue to ask for information from time to time as the project continues. If an owner does not provide this information, the contractor may stop work or, under some circumstances, terminate the agreement. In addition, the owner may not "materially vary" its financial arrangements once made without prior notice to the contractor.

Most owners and lenders do not want to provide extensive or detailed financial information to the contractor, and a contractor likely would not provide such to the lender or owner if the shoe were on the other foot. Although it may be reasonable to receive some assurance that it will be paid, it is inappropriate for the contractor to have intimate knowledge of the owner's and lender's business. Moreover, absent legal requirements to the contrary, a lender generally is not in privity with the contractor and does not owe any duty to the contractor. If the lender supplies information, if may subject itself to direct obligations to the contractor it otherwise would not have had. An owner should strike or substantially modify any provision requiring financial information and should require an acknowledgement that the contractor is not a third party beneficiary of the owner's loan documents.

When a project is financed, some of the questions that counsel to the lender should consider are:

1. Are the contractor and architect (and other consultants) acceptable to the lender?
2. Are the plans and specifications that are referenced in the construction contract the same plans and specifications that the lender has reviewed and approved?
3. Is there a GMP or stipulated sum? Has the lender signed off on the amount? Are there too many allowances or exclusions to the GMP or stipulated sum? Are the unit prices, if any, acceptable?
4. Are the payments to the contractor coordinated with the payments to the owner made under the loan agreement? Is the retainage consistent? Are the requirements of the loan agreement for submissions consistent with those of the construction contract? For example, if the loan agreement requires that a certain form be submitted by the contractor with each draw request, does the construction contract require the contractor to submit such document? How will change orders be funded? What are the conditions to final payment?
5. Are there procedures for obtaining lien releases (or waivers) from the contractors, subcontractors and architect, as permitted by state law? Before release of final payment, must the contractor evidence that it has paid all subcontractors? Is the contractor obligated to pay all lien claimants? May the owner withhold payments to the extent there is a filed or threatened lien?
6. Are any special payment provisions mandated by law being followed?
7. Is the time for performance of the work consistent with the time required under the loan agreement? Are there

provisions for liquidated damages? If so, are they sufficient and enforceable?

8. Must the contractor make claims for delay within a specific period of time? Must the architect approve them? What are the dispute resolution procedures?
9. Should the rights of the contractor and architect to assign their contracts be limited?
10. Is the insurance required of the contractor and architect adequate, and for the lender's benefit?
11. Is the contractor required to obtain all the permits? If hearings are required, are the contractor and architect obligated to participate in all necessary proceedings?
12. Is payment provided for on-site or off-site stored materials? If so, are there protections for the lender?
13. Are there sufficient warranties? Are warranties of subcontractors to be delivered to the owner at the end of the job? Will the warranties also benefit the lender?
14. Are there clear default provisions? Does the lender receive notice and have a right to cure (with longer cure periods)?
15. Is there a contingent assignment of subcontracts and the architect's agreements with its consultants?
16. Is the lender included in the indemnities?
17. Has the lender and its attorney reviewed the general requirements (Division 1) of the specifications?
18. Must the contractor and/or its subcontractors be bonded? Is the lender an obligee under the bonds? Has the lender reviewed the dual obligee rider and the bonds?
19. Have the contractor and architect executed a consent to assignment? Will the contractor and architect perform for the lender? Are there other protections for the lender included?

20. Are the obligations in the architect's agreement consistent with the requirements of the construction contract?
21. Must the architect provide all services necessary to complete the project for basic compensation?
22. Who is inspecting the contractor's work and under what standard?
23. Who reviews shop drawings and is responsible for errors or omissions? Is the contractor responsible for any portion of the design? If so, does that responsibility comply with applicable laws, and is the contractor properly insured?
24. Who coordinates the work of subcontractors, multiple prime contractors and other consultants?

VI. CONCLUSION

The process of documenting a construction project can be a helpful one for all of the participants. Agreements that are carefully reviewed and that are tailored to the particular project force the parties to focus on the issues and to understand the risks. Contracts should allocate risks to the parties that best can control them and should reflect the parties' realistic expectations for the project. A successful project is one where the agreements need never be seen again after they are executed and safely tucked away in the lawyer's files.

2 The Design Professional

Thomas M. Keranen

I. INTRODUCTION

There was a time in the past when the architect enjoyed a significant level of immunity from being sued, as did many of the other traditional professions. In general, only the client had the right to sue the architect, and then, only if there was a breach of the architect's duty of good faith and loyalty to the client. If the architect was found liable, the architect was at risk of surrendering his fee.

Obviously, such is not the case today. Over the years, we have seen substantial and continuing erosion of the traditional immunity of the design professional from suits. Now architects and engineers are exposed to liability for the full consequences of their actions, to their clients with whom they are in privity; to the construction industry which relies on their professional work product, even though there is no privity; and to third parties who allege injury from using the facility. Thus, when undertaking the representation of a design professional in the construction setting, the lawyer must be aware of the full range of potential liabilities faced by the design professional, including the dispute-filled, often litigious nature, of the construction industry.

It has been said that the United States is the most litigious country in the world, and that we have more attorneys per capita than anywhere else. Thus, it perhaps should come as no surprise that litigation has permeated the construction industry, as it has the other facets of our every day lives.

The "litigation explosion" in the construction industry has been a relatively recent trend, with the most acute escalation occurring in the 1970's and 1980's, when interest rates rose sharply. The high cost of money created very large damages on a project that "went bad." As a result of the trend, many contractors, owners and design professionals have experienced more litigation during the past few years then they have during their entire careers.

The present atmosphere is in stark contrast to earlier times "when construction was fun." During the "old days," the atmosphere on the job site tended to be more cooperative, with feelings of trust and camaraderie among the players. There were feelings of helpfulness and teamwork, with a strong interest in making the project successful. Financial margins were larger, so there was "room" to do the little things without requesting extra money. Disputes on the project were normally resolved by informal negotiation at the project level, without resorting to litigation. If there was a formal dispute, it was often resolved by arbitration, which also was less formal and more accessible than it is today.

Industry observers have suggested that the trend toward increased litigation was created by a variety of factors. The industry has grown in size and complexity; the building process is more complicated; contractors tend to be more specialized and do less work with their own forces; the industry is no longer "localized," so participants no longer know each other as well as they traditionally did; economic pressures are far more acute on all of the parties to the process, with smaller financial margins and less margin for error; owners, including governmental owners, tend to be increasingly leveraged; time pressures and accelerated schedules are becoming the norm, thereby increasing job stress and risk on participants; and industry trends towards risk shifting (by industry form contract documents) encourages maneuvering by the parties to reduce exposure which, in turn, tends to create and escalate dispute issues.

Mark Appel of the American Arbitration Association has generally summarized the situation by suggesting that:

> "Construction contracting exists in a largely adversarial atmosphere where collective needs and interests are either subordinated to individual needs or are ignored entirely. The result? Closed lines of communication, mistrust and escalating disputes."

Sealed competitive bidding seems to be a root cause for a large part of the problem. The process—because of its intense competitive nature—minimizes pre-bid communication; eliminates the pre-bid exchange of expertise and information; segregates members of the "project team" (owner, design professional and contractor) into different camps with different interests and objectives for the project; and creates mistrust and misuse of project tools such as the progress schedule. All too often, competitive bidding is the mandate of government, which is not prepared to deal effectively with an active construction project. The weight of the bureaucracy is often in direct conflict with project decision making and timing, thereby distracting from opportunities for project efficiencies. In short, competitive bidding tends to produce a predatory, self-protecting atmosphere, yielding adversarial relationships right from the start.

The effects of these factors, taken together with our society's generally more litigious nature, has caused the traditionally cooperative relationships among the participants to deteriorate, only to be replaced with adversarial, antagonistic ones, "win-lose" attitudes, and general dissension on the job site. The atmosphere has necessitated more lawyer involvement, more claims, increased costs for non-productive effort, and a "legalization" of traditionally informal industry dispute resolution mechanisms.

This has all come at a significant cost to the industry. As Lester Edelman, former General Counsel for the U.S. Army Corps of Engineers once said: "*I believe that quality work is never achieved in an adversarial relationship.*"

Consistent with industry trends, claims against design professionals have grown in numbers during recent years. While there has been some leveling off and decline in the number of claims in the 1990s, the rate of incidence is still much higher than it was 20 or 25 years ago. Much of this trend can be explained by the increased aggressiveness of owners in pursuing claims against the design professional for defective design, cost overruns and delays on their projects. Where the design professional once enjoyed some protection from claims of this kind, traditional legal defenses have eroded, and exposure has increased. Owners have grown frustrated with having to constantly pay for mistakes on their projects, despite the fact that they were not the designer nor did they construct anything. The trend among owners is to hold the design professional and contractor accountable for project performance, including the delivery of the job on time and on budget.

Claims experience would support these observations; approximately half of all claims against design professionals originate with the owner/client. An additional 15% of all such claims originate with the contractor, with about 30% coming from third parties.

As could also be expected, certain types of owner/clients tend to initiate claims more frequently than others. For example, developers are responsible for about 43% of the owner-initiated claims against design professionals, with private owner/users accounting for another 23%. Governmental units account for approximately 18% of owner claims against design professionals. About 43% of all claims are against architects, 26% of claims are against civil engineers, 14% against structural engineers, and 9% are against mechanical/electrical engineers.

The industry has attempted to stem the increase in disputes and to regain the more congenial atmosphere of times gone by. Through the use of "partnering" agreements and non-binding disputes resolution mechanisms (such as structured "step" negotiations, facilitative mediation, dispute review boards and similar devices), the industry has attempted to create better communication, more cooperation, reduced risk and more constructive and efficient dispute resolution.

Despite the outward attempts at creating harmony, some divisiveness has been created by current versions of industry form documents. Leading, commonly used documents, are those published by the American Institute of Architects (AIA), the Associated General Contractors (AGC), and the Professional Engineers in Private Practice (a practice division of the National Society of Professional Engineers, the American Consulting Engineers Council, the American Society of Civil Engineers, and the Construction Specifications Institute), the so-called Engineers Joint Contract Documents Committee (EJCDC). The documents all contain significant "risk shifting" and exculpatory language, intended to "assign" risk of various aspects of the project to one party or another. It goes without saying that each separate form document, while touting fairness and objectivity, has been drafted to generally protect the interests of the drafter.[1]

It is within this complex, hostile atmosphere, with significant exposure at every turn and with each service rendered, that the design professional works. The attorney representing the design professional must have a clear understanding of the industry in order to provide competent advice and consultation. Addressed below are some of the more specific details of that representation.

---

1. As evidenced by a recent article in *Engineering Times*, a publication of the National Society of Professional Engineers, encouraging engineers to use the EJCDC contract documents on their projects, as the forms are protective of the engineer's interests. 23 ENGINEERING TIMES 4 (April 2001).

II. WHAT IS DESIGN AND WHAT DOES THE DESIGN PROFESSIONAL DO?

"Design" has been defined by Webster's Dictionary as the act of creating a purposeful plan or scheme. The person making the plan is, of course, the "designer."

In the context of the construction industry, the "design professional" is an architect or engineer. Architects and engineers are considered "professionals" and are licensed by their states to practice their profession. Under most state occupational codes, the design professional is the person legally responsible for creating the design (the plan or scheme) for the project.

The act of planning and designing the project, is fundamental to developing the project, and the first thing that needs to be done in order to actually build it. Industry lore suggests that a "well planned project is a successful one." Thus, good designers are very valuable commodities. Anyone who can create an aesthetic, pleasing design that functions properly, within the owner's budget and time constraints has certainly made a substantial contribution to our society and is to be highly regarded.

Design is a creative process and good designers are highly respected for their individual talents. Developing the project is often a very exciting and creative time for the owner, since many projects are developed to meet a specific need or want within the community or business organization. "*After long last, we've finally passed the bond issue to build the new school. This will be a monument to our community.*" Thus, the project planning and design process occurs at a time of great expectation, high interest, with much owner enthusiasm. The excitement only escalates during the design process, as design is creative and fun. The project tends to grow in size, feature, and cost. "*Wouldn't it be neat if we could add that . . .*"

The design professional is the leader of the process, and typically comes to the owner after a selection procedure by which the owner selects the designer considered to be the most qualified for the project. Depending on the project, it is often a point of prestige for the owner to be able to select and afford a nationally recognized designer for their project. At the outset, for most projects, the design professional commands the respect of the owner, as one eminently talented and qualified, and having the full capability of delivering a successful project. As such, the design professional occupies a position of leadership for the project, normally working directly with senior management of the owner's organization.

Design professionals like to be in charge of the project, so they tend to readily accept the position of leadership which falls to them. Being in charge allows for the maximum intellectual freedom to develop the design with the least possible restraint.

This makes sense, since design is a combination of art and science. Much of design is an artistic endeavor. Design is not an exact science, in the sense that we cannot simply plug variables into an equation in order to create it. Design does have, on the other hand, aspects to it that are somewhat empirical, such as the design and sizing of structural components or heating and ventilating equipment. Good design requires talent, professional expertise, experience and judgment.

This traditional role of the design professional has historical roots. In times past, the architect was viewed as the "master builder," with total responsibility for and control over all aspects of the project—not only its design, but its construction as well. Architecture was one of the original professions recognized by society.

Over time, the role of the design professional has diminished in scope to the point that design is separate from construction under the traditional methods of project delivery. Current trends in the industry towards design/build project delivery seem to be moving the design professional into a more active role in the construction process once again.

The design process typically starts with making broad determinations regarding the requirements for the project (called a "program"). From the program, preliminary designs are developed demonstrating the general relationships and use of the spaces, aesthetic issues, and other components and features of the project. From there, the design is refined and developed in more detail, culminating in the production of "construction documents." The construction documents represent the final design in the form of drawings and written specifications, which are sufficiently detailed and suitable for bidding and building the project. The contractor then uses the design documents to develop pricing for the work, to submit bids, to develop schedules and work plans, and to actually build the project.

The "design professional" is an architect or engineer, properly licensed and authorized to render designs in the state where the project is located. Under the State of Michigan's Occupational Code, for example, an architect is "... a person licensed under this Act." Thus, merely holding a degree in architecture does not permit a person to be an "architect" in Michigan; to be an architect, a person must first obtain a license to practice architecture from the state.

To obtain a license as an architect or engineer, a person typically must obtain a degree in architecture or engineering from an accredited university; demonstrate character and fitness to practice the profession; and demonstrate a professional competence, usually by serving an internship within the industry, and taking and passing a licensing examination.

Under academic notions, there is little difference in the manner in which the law treats an architect or an engineer. Both are licensed professionals and the law recognizes them as such in applying the professional standards of care and practice. However, in reality, there is a significant difference between the two professions. While both professions create "designs," the subject matter of their respective professional activities differ. The architect tends to deal with buildings, focusing on the spatial concepts and the artistic and aesthetic aspects of the project. Engineers tend to deal with the building's systems and the empirical aspects of the project, such as the sizing and configuration of structural members, or the size and location of the sewer lines. Engineers also have a portion of the industry to themselves in the public works and "civil" works areas, where there is often little involvement by the architect.

At the risk of some over-simplification, and in a very general sense, architects tend to be the "general contractor" of the design process on building projects. Typically, the architect is the lead design professional, and holds the "primary" contract for the design services for the project with the owner. Much like a general contractor, the architect then retains independent consultants, many of whom are engineers, as "subcontractors," to perform related aspects of the design, such as the structural design, the mechanical and electrical designs, and the "civil" or site design. Sometimes the architect will employ consulting architects as subcontractors to handle a specialized aspect of the project. The architect and its consultants each prepare the designs, drawings and specifications for their particular aspect of the project. The architect then receives, coordinates and integrates the individual designs into a final design package consisting of drawings and specifications. We typically see this arrangement in the design of commercial buildings, schools, hospitals, prisons, hotels, casinos and other "architectural" projects of like kind.

A diagram of the contractual relationships on a typical building project is as follows:

On "public works" or "civil construction" projects, the engineer is in the lead, and is the "general contractor" of the design process, instead of the architect. The engineer retains independent consultants as its subcontractors, to perform related or specialized aspects of the design including surveying and mapping, geotechnical evaluation and design,

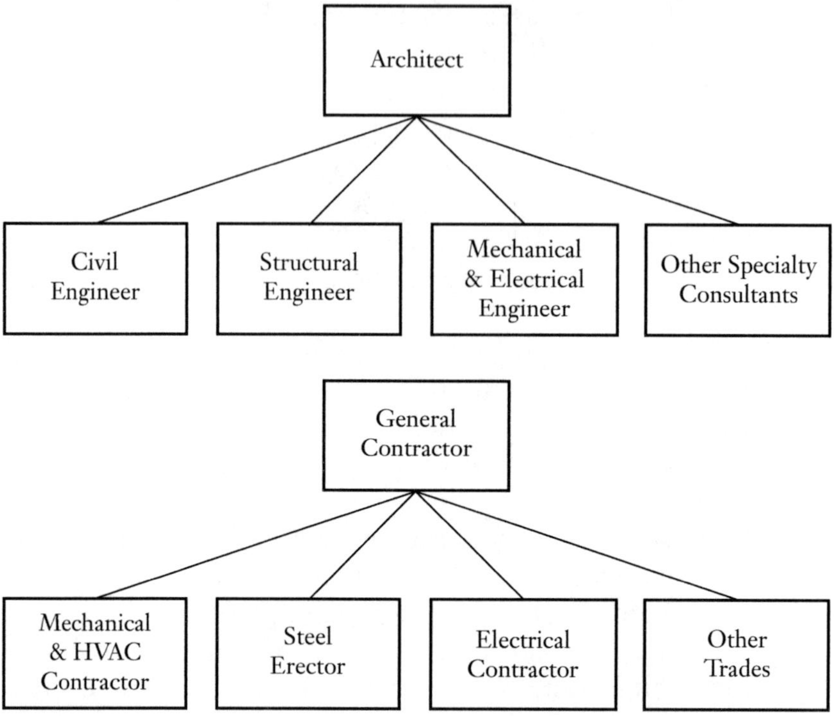

Figure 2.1

structural design, mechanical and electrical design, and so on. Where aspects of these projects involve architectural design, an architect may be employed as a subcontractor to the engineer for that purpose. Projects where this arrangement is typical include highways, runways, bridges, water and waste water treatment plants, sewer and water systems, dams, tunnels, and the wide variety of civil and public works projects.

While the specifics and substance of architect's and engineer's respective practices are different, the fundamental role remains the same: to perform the design function for the project. Thus, it is readily apparent that the design professional's role is pivotal to the project. The design professional is the "first guy in" and takes the lead, on the owner's behalf, much like the captain of the ship, to start the project moving, and to keep it moving from concept through construction.

The fundamental purpose of the design professional's services is to deliver a project that is in accordance with what the owner wants. While the owner's desires for a particular project can appear complex, in the end, the owner's objectives for any project can be categorized in three simple categories:

1. The cost of the project;
2. The time it will take to deliver the project; and
3. The function or purpose of the project.

Most every owner is concerned with how much the project will cost. Obviously, the owner wants to make sure there are sufficient means to pay for the job. It does not matter whether the project is to remodel a kitchen or to construct a new terminal at the airport; cost is almost always an important factor to the owner. In view of the significance of this issue, it stands to reason that the owner expects the design professional to design the project to stay within the budget.

Similarly, the length of time it will require to design and build the project is typically important to the owner. At a minimum, the time allotted to build the project is important to know, so the owner can plan the financial arrangements and cash flow necessary to support the project. Furthermore, often bringing the project on-line by a certain date is critical to the operation of the owner's business.

Finally, the function or purpose of the project is always of fundamental importance to the owner, as it's the very essence of the project. If the owner needs a new hospital, the owner obviously wants the new project to fulfill that need, and function within the established parameters.

The relative priority of the owner's objectives, vis-à-vis each other, might change depending on the type of project, but the owner's objectives always remain the same. If the owner is building a casino, for example, the time for delivery might be the owner's highest priority (as the sooner the casino is open the sooner it creates income), with function being second and cost last. On the other hand, if the project is a new school, funded with a bond issue, delivering the project within budget might be the owner's highest priority (as there may be no additional funds available), with function being second and time of delivery being third. The owner's highest priority for a manufacturing plant might be function (the product cannot be produced without the new plant), with time of delivery being second and cost being third.

It is important for the design professional to fully identify and understand what the owner's interests are in the project, in order to provide successful service (back to the "well-planned project is a successful one" adage). When the design professional takes the seat at the head of the project planning table, it should come as no surprise that the owner expects the design professional to "deliver" the project in a manner that meets the owner's objectives. As a minimum, every owner expects the project to function properly and to be delivered on time, and on budget.

It is further important for the design professional to recognize that each owner has a different level of sophistication in the building process. Some owners are very sophisticated with their own in-house staff of architects and engineers, and have well-established procedures governing their method of designing and building their projects. There are other owners who are entirely unsophisticated and know nothing of the process. A fundamental aspect of the design process is for the design professional to ascertain the relative level of the client's knowledge and to provide the requisite instruction and information sufficient to afford the client the opportunity to fully understand recommendations and to make informed decisions.

Not only is the design professional the general contractor of the design process, but the design professional is usually the owner's representative during construction of the project as well. In the capacity of the owner's representative, the design professional serves as the owner's agent, and provides the point of contact with the contractor and others working on the project. Thus, the design professional provides a variety of services throughout the project.

In addition to providing advice and consultation to the owner/client, and developing and coordinating the design for the project, the design professional is typically called upon to determine the method of project delivery (general contract or construction management) and to assist the owner in the bidding and/or negotiation of the contract with the constructor (general contractor or construction manager). During the construction period, the design professional usually "administers" the work on the owner's behalf by reviewing and approving the contractor's shop drawings and other submittals, making periodic observations and inspections of the work, certifying the contractor's applications for payments; determining and/or certifying when the project has been completed; making interpretations of the design and construction requirements; determining the need for and processing change orders and adjustments to the work; acting upon and assisting with the resolution of claims and disputes and otherwise acting to protect the owner's interests.

In fulfilling this administrative role, the design professional is faced with dual obligations. As a licensed professional, the design professional's first obligation is to protect the public at large, by designing and administering the project to comply with applicable building codes and standards of construction. Simultaneously, under the terms of the business relationship with the client, the design professional is obligated to create a design which meets the owner's objectives, including the owner's program, budget, and time constraints.

III. THE WORK OF THE DESIGN PROFESSIONAL

Although the design professional's role can vary from one project to the next, the design professional's participation on a typical construction project generally occurs in six distinct phases. While the separate phases of activity are now usually defined in the various industry form contracts, they developed historically as being the basic way in which design is developed and then administered during the construction process to assure that the project is built in accordance with the design. The basic phases of service, each of which is discussed below, are programming, schematic design, design development, construction documents, bidding or contract negotiation and construction administration.

A. Programming

The first phase of the design professional's work with the client/owner is to "program" the project. The engineering community calls this activity the "study phase" of the service. During this phase of service, the design professional meets with the owner to determine the owner's needs and objectives for the project. Using an elementary school as an example, the programming process would determine how many students and faculty the school would serve; how many classrooms are needed and how big they should be; what other spaces are needed (gymnasium, cafeteria, administrative offices, etc.) and how big they should be; the site location; utility and service requirements; site requirements; preliminary budget; and schedule requirements. The design professional usually delivers a document, consisting of a written narrative, charts and sketches, to document the program. The owner is normally asked to approve the program in some formal way, prior to moving to the next phase of service.

On an engineering project, the engineer conducts a "study," comparable to the architect's program. In so doing, the engineer is, in essence, developing the program for a public works facility (such as a sewer system) as opposed to a building. Using a sewer system as an example, during the study phase, the engineer will determine the need and size of the sewer system; its general location; its relationship to treatment and other facilities; a cost estimate; and a construction schedule. Often in public works, the study is coordinated with a funding source (such as an EPA grant or USDA funding) in order to meet the criteria for a grant or loan to finance the project. Again, the study is normally published in written form and approved or accepted by the owner before moving to the next phase of service.

B. Schematic Design Phase

The second phase of the design professional's work involves the development of the program into more detailed "schematic" or preliminary plans for the project. The design professional meets with the owner, learning more about the owner's needs and budget for the project, and assisting the owner in the preparation of a more developed description of the project and a more detailed refinement of the original program. Aesthetic issues are developed at this stage, along with more detailed analysis of spatial relationships and other characteristics of the project. As a part of this process, the architect will commence with the preliminary design of the project, that is, the preparation of schematic drawings. Schematic drawings usually illustrate the size and configuration of the building, its "elevation," and its relationship to other components of the project, such as the site layout, parking lots, and the surrounding environment.

During the schematic design phase of the work, the design professional will develop a preliminary estimate of the probable construction cost for the project and prepare and review the preliminary schedule for design and construction. Also during this phase, alternative approaches to the design of the project may be introduced to the owner, and basic design decisions are made, such as whether the building should be constructed of structural steel or reinforced concrete, and the style of the facade of the building. At this stage, the architect's consultants are usually involved in making preliminary designs and decisions regarding their areas of the project.

On an engineering project, this phase of service is often referred to as "preliminary engineering" and consists of similar activities by the engineer. Topographic surveys are done of the project area; investigations of existing utilities, environmental and geotechnical conditions are made; more detailed engineering calculations are performed to determine the capacity, size, service area, and location (route selection) for the facility; and further cost estimates and schedules are created. Preliminary drawings, calculations and specifications are normally prepared to document this aspect of service.

C. Design Development Phase

Once the owner's program, budget and schematic design of the project have been established and approved, the design professional embarks upon the more detailed development of the design. During the design development phase, more detailed design development drawings and specifications are prepared, which describe in significantly more detail the size and character of the project, including its architectural, structural, mechanical, and electrical systems and other features. Throughout the design development phase, the architect communicates with the owner periodically, and refines the design based upon the owner's observations, comments, input, directions and approvals. Budgetary and other constraints play an important part in this process, as it is a typical expectation of the owner that the project will be designed to "fit" the budget.

D. Construction Documents Phase

When the design development documents are accepted and approved by the owner, the final design of the project is embodied in the construction documents, which will be used for the actual construction of the project. The construction documents provide the information necessary to construct the project, and provide the various details, dimensions and technical information relevant to the project. The construction documents are normally delivered in a package consisting of the project manual and the construction drawings. The project manual, commonly referred to as the "specifications," contains the technical specifications for the work, together with the "front end" general conditions and supplementary conditions, which govern the contractor's business relationship with the owner. The technical specifications are often organized in accordance with the Construction Specifications Institute (CSI) format, an industry standard, and contain divisions for the various categories of work (plumbing, electrical, roofing, carpentry, and so on).

The design drawings are generally organized according to disciplines. In a given set of construction drawings, one typically finds a number of architectural drawings depicting the configuration and architectural components of the project (labeled as the "A" sheets, e.g., A-1, A-2, A-3, etc.), structural drawings depicting the structural design of the project (e.g., S-1, S-2, S-3, etc.), mechanical drawings depicting the plumbing, HVAC and other mechanical systems for the project (e.g., M-1, M-2, M-3, etc.), electrical drawings depicting the electrical equipment, wiring diagrams and electrical systems (e.g., E-1, E-2, E-3, etc.) and civil drawings illustrating the site layout, grading, paving, drainage, sewers, water lines, and similar requirements for the project which are generally external to the building itself (e.g., C-1, C-2, C-3, etc.).

At this stage, the design professional is effectively drafting the requirements for the contract between the owner and the contractor. The design professional's work product, consisting of the drawings and technical specifications, defines the "scope" of the contractor's work on the project. The "front end" of the standard form contracts defines the business terms of the relationship (see AIA Document A-201, General Conditions, for example). A typical provision of the general conditions of the construction contract obligates the contractor to construct the project in "strict accordance" with the contract documents, including the drawings, technical specifications and the other conditions of the contract.

In this context, it is important to note that nearly every aspect of the design professional's day-to-day service on a project has legal significance. Accordingly, aside from an analysis of the written contracts between the parties, perhaps the single most important point of departure for a lawyer's assessment of the design professional's liability is a detailed analysis of the project manual and contract drawings; their importance in assessing the relative rights of the parties to a dispute cannot be understated.

E. Bidding or Negotiation Phase

Once the construction documents for the project have been prepared and approved by the owner, the owner, most often with the design professional's assistance, will use them to solicit proposals or competitive bids from contractors for the construction of the project. During the bidding and negotiation phase, the design professional may be called upon to assist the owner in comparing and/or evaluating the bids or proposals received from the various contractors, the qualifications of the bidders to perform the work, and to make recommendations as to which contractor the owner should award the work to.

F. Contract Administration Phase

Once the contractor is selected and construction starts, the design professional typically will undertake the contract administration phase of the work, providing certain administrative functions defined in his contract to assist the owner through the actual construction of the project. During the contract administration phase of the project, the design professional serves as the owner's representative, and point of contact with the contractor. The design professional usually provides services to evaluate the progress and quality of the completed work; to ascertain if the work is being performed in accordance with contract and code requirements; and generally, to try to protect the owner's interests in the project.

As part of the construction administration service in the typical project, the design professional makes periodic site visits to observe the progress of the work; attends job progress meetings; reviews and approves the contractor's shop drawings and submittals; reviews and certifies pay applications; provides advice and consultation to the owner and contractor concerning questions and design issues; determines the need for and processes change orders revising the work; certifies substantial and final completion of the work; reviews and responds to claims and disputes; and develops punch lists and project close-out documentation.

On a public works project site, the engineer is traditionally more actively involved in the day-to-day on-site administration of construction than the architect is on a building project. For example, engineers often have a complement of staff on-site every day, and they perform certain activities which, on a building project, tend to be the contractor's obligations. In this model, the engineer typically provides services such as construction staking to locate the work; field inspection of the work as it is being constructed; and quality control services, in addition to performing the other services identified above. To some extent, even the normal services are performed differently. For example, on the building project the architect normally reviews and certifies a pay application submitted by the contractor. On the public works project, the engineer develops a "pay estimate" based on the engineer's inspection of work performed, independent from any invoice submitted by the contractor.

The majority of claims against design professionals arise out of the actual construction process, and include claims of error or omission in the design, as well as the professional's conduct during the contract administration phase. It is important to note that many design professionals view the construction administration service as having less importance than the design process. Under most industry contracts, 80% to 90% of the fee has been earned by the time the design is put out for bids, and the lead professionals on the designer's staff are on to other projects. Construction administration is left to the professional's construction staff, who often have less experience and status within the firm. The designers, being busy with new work, have little time to spend on administering construction or working out details.

The contrary is true for the contractor, who hopefully will staff the project with its best people, since maintaining good production is key to a profitable job for the contractor. Thus, sometimes there is a mismatch between the relative quality of staff between the contractor and design professional. Since this is the place and time most claims arise, the mismatch in staffs can have negative effects on the owner's and design professional's ability to maintain good job progress and resist claims by the contractor.

IV. PRACTICAL CONCERNS AND ISSUES FOR THE DESIGN PROFESSIONAL

Before turning to a discussion of the design professional's legal rights and responsibilities in relation to the owner, general contractor, and third parties, an overview of the practical considerations facing the design professional is warranted.

As noted above, architects and engineers, as licensed professionals, have an obligation not only to their clients, but also to the public at large, to design projects which are safe and suitable for their intended use. A common difficulty for the design professional is to develop a design for the project that will satisfy the owner's requirements (such as the owner's budget limitations) while still meeting the governing codes and other public requirements. Many claims against design professionals originate in this struggle. Although designers are often faced with allegations that their design does not go far enough, they may also be criticized for being overly conservative by failing to provide a more economical design.

Another practical consideration facing design professionals is that they almost certainly rely on referrals and repeat business from their clients. Although they are required to, and typically do, maintain independence in their professional judgments, designers often have to choose between the demands of the client to provide their services at the least cost, on the one hand, and their conflicting professional responsibilities to protect the public, on the other. The challenge to a designer's professional independence is perhaps best illustrated in the circumstance where the design professional is called upon to render an interpretation and opinion of the requirements of the contract documents (which the design professional drafted) when a dispute between the owner and contractor erupts over what the requirements are.

Another frequent practical dilemma faced by design professionals is delivering an acceptable set of contract documents for the project, within the fee negotiated with the owner, and still making a profit. Design professionals are constantly under pressure to keep current with the technical advances in the construction industry, including the latest changes in design approaches and systems. Finding an appropriate balance between devoting resources and personnel to a project and maintaining profitability is a recurrent challenge. A frequent source of litigation against design professionals is the lack of "coordination" in the contract documents and drawings among the various design disciplines. For example, the mechanical drawings have not been coordinated with the architectural drawings which show mechanical features at locations different than those shown on the architectural plans. Although the architect is responsible to the owner for the entire design on a project, including the work of its subcontractors and consultants, there is frequently little time or money available in the fee for a detailed examination and coordination of the consultants' work product by the architect.

A related concern of design professionals is assuming full responsibility for the design and contract administration of a project under contract terms which limit the scope of services, and the fee for services. Owners often want the design professional to "protect" their interests, but often are unwilling or unable to pay for a full compliment of available services to protect those interests. This is most often evidenced at the contract administration phase of the work where the owner wants the design professional to inspect the construction as it progresses. Rather than pay for a full time inspector, the owner may insist that the architect limit site visits to periodic observations of the work, in an effort to lower the overall fee. While design professionals are often willing and capable of providing an expansive array of services which can assist the owner and facilitate the achievement of the owner's objectives, the owner's budgetary constraints may preclude the design professional from providing such services.

On a personal level, a frequent source of frustration for design professionals is the owner's limitations on creative approaches to design, which are dictated by financial or other restraints. It is a rare occurrence for designers to be given the latitude fully to pursue creative design approaches, especially on matters of aesthetics, without severe owner-imposed limitations.

Having considered a few of the practical concerns facing the design professional on a construction project, our discussion now turns to the specific relationships and legal responsibilities of the design professional with the owner, the general contractor, members of the public, and the design professional's consultants.

V. THE DESIGN PROFESSIONAL'S RIGHTS AND RESPONSIBILITIES IN RELATION TO THE OWNER

The design professional's rights and responsibilities in relationship to the owner are generally governed by the terms of the agreement between the owner and the design professional.[2] However, in addition to the terms of the agreement, the design professional is always required to render professional services in accordance with the standard of care applicable to members of the profession. A design professional generally has the legal duty to exercise that degree of care and skill which a reasonably prudent designer would use under the same or similar circumstances.[3] It is noted that many of the industry form contracts (the AIA Document B-141, for example) adopt the professional standard of care as the contractual standard under which the designer's services will be performed.

In undertaking responsibilities on a project, the design professional is generally entitled to rely on information furnished by the owner regarding the project's requirements, characteristics, surveys, conditions and other specifics.[4] This notion arises from risk assignment language found in the AIA and EJCDC documents that obligates the owner to furnish owner-held information, and accept responsibility for it. This general proposition usually applies unless the design professional becomes aware of errors or inaccuracies in information supplied by the owner, or unless the design professional's contract provides otherwise. It is not uncommon, however, for the owner's information (for example, the drawings for an existing building) to be outdated. The owner expects that as part of the designer's service, the outdated drawings thus provided will be checked and updated. The standard contract language would seem not to require this activity, illustrating an example where the owner's expectations of service is at variance with the designer's. This leads to the legal dilemma: *who is responsible if the outdated drawings are never checked or updated,*

---

2. Getzschman v. Miller Chem. Co., 232 Neb. 885, 443 N.W.2d 260 (1989).

3. Friendship Heights Assoc. v. Viastimil Koubek, AIA, 573 F. Supp. 100 (D. Md. 1983), *aff'd in part, rev'd in part,* 785 F.2d 1154 (4th Cir. 1986); Perlmutter v. Flickinger, 520 P.2d 596 (Colo. App. 1974); Crockett v. Crothers, 264 Md. 222, 285 A.2d 612 (1972).

4. See, e.g., art. 4.9 of AIA Document B-141, *Standard Form of Agreement Between Owner and Architect* (1987 ed.).

and a differing site condition is subsequently encountered by the contractor, arising directly from the failure to update the drawings, thereby obligating the owner to make an adjustment to the contract price and/or time?

In addition to relying upon information furnished by the owner, a design professional is also entitled to expect reasonable compensation for its professional services, in accordance with the terms of the agreement with the owner. In the absence of an express agreement on compensation, the design professional will usually be entitled to the fair and reasonable value of the services.[5]

The corollary to the design professional's rights and reasonable expectations are the reciprocal obligations and responsibilities to the owner. Although design professionals do not guarantee the results of their work, the design professional will be liable for errors and omissions in the plans and specifications if they result from the professional's negligence,[6] or if they constitute a failure by the design professional to meet contractual obligations.[7] In addition to their own errors and omissions, the design professional is also responsible to the owner for the errors and omissions of their consultants, usually under a contract or *respondeat superior* theory.[8]

A frequent accusation of failure on the part of the design professional, relating to errors or omissions in the contract documents, is the professional's failure to properly coordinate the work of the consultants. As suggested earlier, in addition to improper coordination, the professional can be held liable for either over-designing or under-designing a project if that design fails to meet the requirements of the

---

5. Whitney v. Lea, 367 A.2d 683 (Vt. 1976).

6. Kellogg v. Pizza Oven, Inc., 402 P.2d 633 (Colo. 1965).

7. Irondequoit Bay Pure Waters Dist. v. Nalews, Inc., 123 Misc. 2d 462, 472 N.Y.S.2d 842 (1984).

8. Atlantic Nat'l Bank v. Modular Age, Inc., 363 So. 2d 1152 (Fla. Dist. Ct. App. 1978), *cert. denied*, 372 So. 2d 466 (Fla. 1979); Johnson v. Salem Title Co., 425 P.2d 519 (Or.1967).

professional's contract, or the standard of care required for the project.[9]

Another frequently-alleged error subjecting the design professional to liability is the existence of omissions or "holes" in the plans or specifications. A simple illustration of this concept is where the mechanical drawings call for the installation of a certain piece of equipment, and the electrical drawings fail to indicate the required power service for the equipment. While the documents imply the installation of the electrical service to operate the mechanical equipment, taken literally, there is no electrical service shown, so the mechanical equipment would be installed without the electrical power to operate it.

Aside from errors or omissions in the design or construction documents, the design professional's second greatest exposure on a construction project often lies in potential deficiencies in the performance of the construction administration phase services. Design professionals are often accused of failing to reject work which is deficient or not in accordance with the requirements of the contract documents;[10] failing to report deviations from the plans and specifications, or from the requirements of the applicable building codes;[11] and improperly rejecting or delaying action on the contractor's shop drawings or other submittals.[12]

While the general contractor is generally entitled to decide its own means and methods of construction, design professionals are increasingly exposed to claims of interference with the contractor's completion of the work. In the performance of the construction administration services, the professional is expected to observe the progress and quality of the work in order to determine whether the work com-

---

9. Donnelly Constr. Co. v. Oberg/Hunt/Gilleland, 677 P.2d 1292 (Ariz. 1984); Fox v. Stanley J. How & Assocs., Inc., 309 N.W.2d 520 (Iowa Ct. App. 1981).

10. Skidmore Owings & Merrill v. Connecticut Gen. Life Ins. Co., 197 A.2d 83 (Conn. Ct. App. 1963).

11. Miller v. De Witt, 226 N.E.2d 630 (Ill. 1967).

12. 530 East 89 Corp. v. Unger, 373 N.E.2d 276, (N.Y. 1977).

plies with contract requirements and whether contractor's applications for payment should be approved. A typical source of exposure to the professional, particularly in projects where the contractor has defaulted under the contract, is the accusation that the professional improperly certified the contractor's applications for payment while overlooking the contractor's default.[13]

Another significant source of claims against design professionals is the added cost necessary to complete or correct deficiencies on a project occasioned by the design professional's negligence. Such deficiencies are often alleged to have resulted from errors and omissions in the design, which led to additional work for the contractor, as well as claims for delay or disruption damages from the contractor and/or owner. Damages against the design professional are usually measured by the actual cost of the additional work necessary to correct the errors and omissions in the design, and/or any additional costs occasioned by the resultant delay or disruption to the construction process. However, when the cost of repair or correction is so great as to create an unreasonable economic waste, the diminution in value of the project as a whole may be used as the measure of damages.[14]

VI. THE DESIGN PROFESSIONAL'S RESPONSIBILITIES IN RELATION TO THE CONTRACTOR

As a general rule, the design professional's liability on a construction project, under contract law principles, extends only to the party with whom the professional has a contract, that is, the owner.[15] Although contractors often attempt to

---

13. Barrett, Daffin & Figg, Architects-Planners-Eng'rs, Inc. v. McCormick, 362 So. 2d 966 (Fla. Dist. Ct. App. 1978), *cert. denied*, 368 So. 2d 1362 (Fla. 1979); Browning v. Maurice B. Levien & Co, 262 S.E.2d 355 (N.C. Ct. App. 1980).

14. Grossman Holdings Ltd. v. Hourihan, 414 So. 2d 1037 (Fla. 1982).

15. Steven G.M. Stein, 1 *Construction Law*, at 5-181 (1993); Frank D. Wagner, Annotation, *Tort Liability of Project Architect for Economic Damages Suffered by Contractor*, 65 A.L.R.3d 249, 252 (1975).

pursue contract-based claims against the professional using a third-party beneficiary theory, most modern day owner/design professional agreements specifically negate any third-party beneficiary entitlement for the contractor.[16] In those instances where the third-party beneficiary analysis is applicable, the success of the contractor's claims typically turns on whether the contractor's reliance on the design professional's performance was reasonably foreseeable to the design professional.[17]

A more prevalent theory of liability advanced by contractors against the design professional is the professional's liability in tort (malpractice) for failing to exercise the requisite standard of professional care. The tort liability of design professionals to contractors and other third parties, however, varies from one jurisdiction to the next. Many jurisdictions preclude a contractor from recovering against the design professional in tort, where the contractor has only suffered economic losses, and where there is no privity, under the economic loss doctrine. Jurisdictions, which continue to apply this "economic loss doctrine" as a bar to contractor's recovery, include Virginia, Florida, and Illinois.[18] Other jurisdictions, including Maryland and Nevada, have relaxed the economic loss doctrine to permit recovery for purely economic losses depending upon the nature and severity of the risk of damages occasioned by the design professional's negligence.[19]

The design professional's liability to a contractor for negligence can arise from errors and omissions in the drawings and specifications, including those prepared by the consultants.[20] An example of a court upholding a contractor's

---

16. *See*, e.g., art. 9.7 of AIA Document B-141 (1987 ed.).

17. *Note* 92 HARV. L. REV. 1076 (1979).

18. Blake Constr. Co. v. Alley, 353 S.E.2d 724 (Va. 1987); Redarowicz v. Ohlendorf, 441 N.E.2d 324 (Ill. 1982); Florida Power & Light Co. v. Westinghouse Elec. Corp., 510 So. 2d 899 (Fla. 1987).

19. Council of Co-Owners Atlantis Condominium, Inc. v. Whiting-Turner Contracting Co., 517 A.2d 336 (Md. 1986); Local Joint Executive Bd., Culinary Workers Union, Local No. 226, 651 P.2d 637 (Nev. 1982).

20. Atlantic Nat'l Bank v. Modular Age, Inc., 363 So. 2d 1152 (Fla. Dist. Ct. App. 1978); Johnson v. Salem Title Co., 425 P.2d 519 (Or. 1967).

legal standing to sue the design professional, despite the lack of privity, can be found in the Michigan case *Bacco Construction Company v. American Colloid Company, D.J., et al.*[21] *Bacco* involved claims by the contractor for the construction of a municipal waste water lagoon and irrigation system against the designer and engineer for the project, McNamee, Porter & Seeley. There was no privity between Bacco and McNamee. The court determined that, notwithstanding the lack of privity, the contractor could maintain an action against the project engineer, because the engineer's failure to make proper calculations and specifications for the project could have foreseeably been seen to create the risk of harm to the contractor responsible for using the specifications to construct the job. The court held that McNamee's exposure included the risk of financial hardship on the contractor of having to cure the defects caused by McNamee's defective specifications.

Another frequent theory of liability advanced against design professionals by contractors is negligence in the supervision and/or inspection of the construction work. The design professional's duty to observe the work is generally governed by the owner/design professional agreement. Absent the undertaking of a specific obligation, the design professional generally owes no duty to the contractor to supervise the contractor's work.[22] Generally, the duty to periodically observe the work only creates a duty upon the design professional to perform inspections in accordance with the applicable standard of care. It does not create a duty to assure the protection of the safety and welfare of workers on a construction site.[23] Although the professional's contractual obligations may be limited to making periodic observations or inspections of the work, the professional can be held liable for failing to observe obvious deficiencies in critical

---

21. Bacco Construction Co. v. American Colloid Co., D.J., 384 N.W.2d 427; (Mich. Ct. App. 1986).

22. Columbus v. Clark-Dietz & Assocs.-Eng'rs, Inc., 550 F. Supp. 610 (N.D. Miss. 1982), *app. denied*, 702 F.2d 67 (5th Cir. 1983).

23. Miller v. De Witt, 226 N.E.2d 630 (Ill. 1967).

aspects of the work,[24] or for negligently failing to reject defective materials and workmanship.

Another common exposure for the design professional to the contractor, on a tort theory of recovery, is allegations of negligence against the design professional for refusing or failing to act promptly or properly on the contractor's submittals.[25] Contractors will usually attempt to rely upon the professional's "approval" of submittals, even where those submittals may deviate from the requirements of the contract documents. Design professionals are often called upon to render a judgment as to whether equipment being proposed by a contractor is comparable or "equal" to that which is specified. Some professionals have been found liable for enforcing overly restrictive specifications without a rational or scientific basis to support their position.[26] Furthermore, where a professional does grant approval to a contractor's submittal and later reverses himself, he may be found liable for the additional costs incurred by the contractor as a result of its subcontractor's work done pursuant to the earlier approval.[27]

Generally, the design professional owes no duty for scheduling or coordinating the construction of the project since the means and methods of construction are generally within the province of the contractor.[28] However, if the design professional does undertake to assist or direct these duties, the professional may be responsible if he does so negligently.[29]

---

24. Pastorelli v. Associated Eng'rs, Inc., 176 F. Supp. 159 (1959).

25. E.C. Ernst, Inc. v. Manhattan Constr. Co., 551 F.2d 1026 (5th Cir. 1977), *cert. denied*, 434 U.S. 1067 (1978); Gurtler, Hebert & Co. v. Weyland Machine Shop, Inc., 405 So. 2d 660 (La. Ct. App. 1981), *cert. denied*, 410 So. 2d 1130 (La. 1982).

26. Waldinger Corp. v. Ashbrook-Simon-Hartley, Inc., 564 F. Supp. 970 (C.D. Ill. 1983), *aff'd in part, remanded in part*, 775 F.2d 781 (7th Cir. 1985).

27. Detweiler Bros., Inc. v. John Graham & Co., 412 F. Supp. 416 (E.D. Wash. 1976).

28. *See* art. 2.6.6 of AIA Document B-141 (1987 ed.).

29. Peter Kiewit Sons' Co. v. Iowa Southern Utilities Co., 355 F. Supp. 376 (S.D. Iowa 1973).

VII. THE DESIGN PROFESSIONAL'S RESPONSIBILITIES IN RELATION TO THIRD PARTIES

As with contractors, the design professional's responsibility to third parties is generally premised on either a third-party beneficiary theory, or on the basis of tort duties resulting from the professional's conduct.[30] The design professional's liability for deficiencies in the design and resultant damages to third parties often survives long after completion of the project, in spite of the fact that the work may have been completed and accepted years earlier by the owner, and even approved by the controlling building officials.[31] Many states have adopted statutes of repose, which in essence place statutory time limitations on when causes of action for damages arising out of defects in the design or construction of a project will accrue.[32] Unlike traditional statutes of limitations, which bar recovery on a cause of action, statutes of repose generally provide that no cause of action will accrue after a specified period of time has passed from the date that the building was completed or first available for its intended use, sometimes measured by the date the certificate of occupancy was issued.

Questions of the design professional's liability to third parties often arise when the supervisory responsibilities exceed mere observation of the progress of the work. Determining whether an injured party can recover against the design professional often depends upon the contractual undertakings of the professional and the professional's specific conduct at the scene of the incident.[33] For instance, for the professional to be liable to a worker injured on a construction site, the professional generally must have assumed some duty to supervise the method or manner of the performance of the

---

30. A.R. Moyer, Inc. v. Graham, 285 So. 2d 397 (Fla. 1973).

31. Totten v. Gruzen, 52 N.J. 202, 245 A.2d 1 (1968).

32. *See, e.g.*, MD. CODE ANN., Cts. & Jud. Proc. § 5-108(b) (1989); PA. STAT. ANN., tit. 42, § 336 (1993); TEX. CIV. PRAC. & REM. CODE ANN., § 16.009 (1993); VA. CODE ANN. 8.10-250 (1993).

33. Hanna v. Huer, Johns, Neel, Rivers & Webb, 662 P.2d 243 (Kan. 1983).

actual construction work, or may have observed the unsafe condition and failed to act.[34] Thus, where the professional has knowledge of a hazardous condition and fails to act appropriately upon that knowledge, the professional can be held liable for personal injuries that are occasioned by the hazardous condition.[35] Similarly, if the design professional has the right to insist upon safe and adequate construction methods and knew or should have known of a hazardous condition, he can be liable for failing to stop the work until the unsafe condition has been remedied.[36] Knowingly permitting a departure from the requirements of the contract documents, which results in a defect in the construction project and resultant personal injury, can also subject the design professional to liability.[37]

VIII. THE DESIGN PROFESSIONAL'S RESPONSIBILITIES IN RELATION TO ITS CONSULTANTS

The liability of design professionals for the consultants employed as their subcontractors under the primary owner/professional agreement is generally governed by contract and agency principles. While the design professional is held liable for errors and omissions in the work of the consultants,[38] the professional may generally rely on the consultants' work product and thereby be entitled indemnification for such liability, usually as a matter of contract or as a matter of law.[39] Although the design professional maintains overall responsibility to the owner and others for the project's entire design, the professional is entitled to rely on the professional

34. Cutup v. Lucky Stores, Inc., 22 673, 325 A.2d 432 (Md. Ct. App. 1974); Krieger v. J.E. Greener Co., 382 A.2d 1069 (Md. 1978).

35. Caldwell v. Bechtel, Inc., 631 F.2d 989 (D.C. Cir. 1980); Geer v. Bennett, 237 So. 2d 311 (Fla. Dist. Ct. App. 1970).

36. Miller v. De Witt, 226 N.E.2d 630 (Ill. 1967).

37. Potter v. Gilbert, 130 A.D. 632, 115 N.Y.S. 425, *aff'd*, 90 N.E. 1165 (N.Y. 1909).

38. Johnson v. Salem Title Co., 425 P.2d 519 (Or. 1967).

39. Alisal Sanitary Dist. v. Kennedy, 4 Cal. Rptr. 379 (1960); Shea v. Bay State Gas Co., 418 N.E.2d 597 (Mass. 1981).

skill and judgment of the consultants, and is entitled to expect that their proper care and judgment will be exercised. Naturally, the design professional is obligated to pay for the services and work of the consultants, in accordance with the terms of their agreement, or in the absence of a specific agreement, on the basis of the reasonable value of the services.

In addition to assisting in the development of the construction documents for the project, the consultants are usually called upon to assist in the construction administration of the project, performing site inspections for those parts of the work relating to their field of expertise, and by reviewing contractor submittals which relate to their design or discipline. For example, a structural engineer may be asked to visit the site to observe the placement of reinforcing steel in the concrete of the building. As licensed professionals, the consultants owe a responsibility not only to the design professional with whom they have a contract, but also to the public at large on whose behalf and for whose protection they are required to exercise their professional judgment.

IX. CONCLUSION

This chapter represents an abbreviated overview of a large, complex subject matter. Design is a difficult job with many pitfalls for the wary, as well as the unwary. Representing the design professional is, perhaps, even more difficult, because the designer is often oblivious to its own legal rights and responsibilities and fails to recognize the resulting exposure to liability for claims. Hopefully this chapter will introduce the reader to a few of the relevant legal topics worthy of further research and mastery as part of the representation of the design professional.

3 Role of the Prime Contractor

John T. Clappison

I. INTRODUCTION

Although the past few years have seen significant changes in the ways in which owners contract for the design and construction of their projects, the prime (or general) contractor remains a cornerstone of the construction industry. In a nutshell, the prime contractor is responsible for transforming a set of plans and specifications into a tangible structure, suitable for the owner's intended use. In order to be successful, a prime contractor must, above all, be an adept manager: one who is able to schedule the work, coordinate subcontractors, procure materials, and maintain cash flow in such a way that the project can be completed on time and within budget. This is no easy task. Regardless of the size or complexity of the project, the path between bid submission and project completion is fraught with obstacles that must be overcome if the prime contractor is to be successful. These obstacles include poor design drawings, unforeseen site conditions, labor strikes, bad weather, unreasonable owners, uncooperative subcontractors and many other things.

With this in mind, the following discussion is intended to provide a brief overview of the prime contractor's traditional role in the construction process.

II. VARIATIONS IN CONSTRUCTION PROJECTS: PUBLIC VS. PRIVATE

Construction projects fall into one of two general categories: public and private. From a legal standpoint, this distinction is significant. Public projects are subject to a number of different statutory and regulatory schemes, including competitive bidding laws, payment and performance bond statutes,[1] federal and state false claims acts,[2] prompt pay legislation, preference programs for disadvantaged business enterprises,[3] special procurement regulations, and special procedures for filing and prosecuting claims.

Private projects, though subject to much less regulation, nevertheless have their own set of rules. Unlike public projects, private projects are subject to construction lien laws, which vary from state to state and often distinguish between residential and commercial construction.[4] In addition, residential construction is frequently subject to con-

---

1. *See* chapters on "Role of Surety" and "Payment Provisions" for further discussion on lien and bond issues.

2. *See* Pages, *The Civil False Claims Act—Legislation in Turmoil*, CONSTR. LAWYER 42 (Jan. 1999).

3. *See* Steven R. Schooley & Michael W. Andrew, Jr., *The Devil in Devolution: State and Local Preference Programs*, CONSTR. LAWYER 18 (Oct. 1996).

4. *Compare* MICH. COMP. LAWS ("MCL") 570.1108 (specifying notice of commencement requirements for nonresidential construction) *with* 570.1108a (specifying notice of commencement requirements for residential construction).

sumer protection laws and their exemplary damages and/or attorney's fees provisions.

These lists are by no means complete, and the laws governing many, if not most, of these issues vary from federal to state, state to state, and city to city. Therefore, the legal practitioner is cautioned to be cognizant of the specific laws that govern a particular project.

III. PROJECT DELIVERY SYSTEMS: DIFFERENT WAYS TO DO THE WORK

A. The Traditional "Design-Bid-Build" Model

The traditional "design-bid-build" model for construction projects involves a tripartite relationship between the owner and the design professional on the one hand, and the prime contractor on the other. Under this model, the owner contracts with the design professional (the "A/E") to design the project, assist the owner in bidding the project and serve as the owner's agent in administering the contract during construction.

The owner then contracts separately with a prime contractor for the actual construction of the project. While the prime contractor may, and typically does, subcontract with numerous specialty trades to perform various components of the work (e.g., electrical, mechanical, plumbing, painting and carpentry), the prime contractor remains solely responsible to the owner for constructing the project.

Because the prime contractor and A/E lack contractual privity under this "traditional" model, the prime contractor has no direct contract-based cause of action against the A/E for design errors. Historically, recovery of losses incurred as a result of poor design has been based upon claims against the owner for breach of the implied warranty of the adequacy of the plans and specifications, a concept first articulated in the seminal case of *United States v.*

Spearin, 248 U.S. 132; 39 S. Ct. 59 (1918).[5] More recently, however, courts have begun to permit direct tort-based claims by prime contractors against design professionals for purely economic injury attributable to design defects, despite the traditional defenses of lack of privity and the economic loss doctrine.[6]

B. Alternative Project Delivery Systems

Over the past few decades, the construction industry has adopted alternative forms of project delivery systems, some of which have altered the traditional role of the prime contractor. Several of the more popular alternatives are discussed below.

1. Design-Build Contracts

Design-build contracts offer one-stop shopping to an owner; the owner enters into a single contract with one entity for both the design and the construction of the project. The design-builder may be a single entity having both construction and design capabilities in-house, a joint venture between a designer and a builder, a contractor who subcontracts the design work, or a designer who subcontracts the construction work. Under any of these scenarios, the premise of design-build is that, with the prime contractor and designer on the same team, they

---

5. Under the *Spearin* doctrine, the owner impliedly warrants that the contractor will be able to perform the contract successfully if it follows the specifications (i.e., that the plans and specifications are free of defects). The doctrine applies only to "design" specifications, which instruct the contractor on what materials to use and how they are to be assembled, as opposed to "performance" specifications, which simply specify a performance level that the end product is to achieve, leaving it to the contractor to determine how to achieve the performance level. *See* Lenderman, *The Spearin Doctrine: It Isn't What It Used to Be*, CONSTR. LAWYER 46 (Oct. 1996).

6. *See* V. William Ernstrom & Pamela L. Ryan, *Expanding Negligence Liability of Design Professional to Contractors*, CONSTR. LAWYER 11 (Jan. 1988).

are better able to address constructibility issues during design, thereby lowering the overall cost of the project and shortening the duration between project design and final completion.

The biggest impact on a prime contractor which assumes the role of design-builder is the assumption of design risk, which remains with the A/E under the traditional design-bid-build scenario. In addition, design-build contracts usually require the design-builder to commit to a guaranteed maximum price for the overall cost of the project at a time when the project is little more than a concept.

One of the primary criticisms of the design-build delivery system, from the owner's standpoint, is the concern that the owner's interests are not adequately protected when the designer and the contractor are one and the same. However, this concern appears to have done little to stem the rising popularity of design-build contracts.[7]

2. Construction Management Contracts

Although perhaps not technically a separate delivery system, another variation on the traditional model involves the increased use of construction management contracts. Under this scenario, the owner contracts with an A/E to design the project. But instead of having the A/E administer the construction contract on the owner's behalf, the owner contracts separately with a prime contractor to serve as a construction manager ("CM") to oversee project construction.

The role of the prime contractor as CM has several possible variations. Common to each is the fact that, as a CM, the prime contractor rarely self-performs any of the work, but rather supervises and manages the work performed by others. The variation comes with respect to the relationship

---

7. *See* Mark C. Friedlander, *A Primer on Industrial Design/Build Construction Contracts*, CONSTR. LAWYER 3 (April 1994); Neil S. Haldrop & Mason A. Goldsmith, *Design-Build in the Public Sector*, CONSTR. LAWYER 38 (Oct. 1998).

between the CM and the trades performing the work, and whether the CM serves solely as the owner's agent, or as a constructor. Under the CM-as-agent scenario, the owner contracts directly with the trades, and the CM serves strictly as the owner's agent in overseeing the construction—much in the way the A/E does under the traditional model. The CM-as-constructor scenario involves the CM contracting directly with the trades and assuming more of a traditional prime contractor role—as well as assuming the risk of completing the project on time and within budget. A third variation involves the owner contracting with a single prime contractor to perform the work, while maintaining separate contracts with an A/E for design, and a CM for contract administration.

The advantage to using a CM is typically realized on large, complex construction projects that require significant managerial emphasis and control. The downside is that use of a CM, particularly in the CM-as-agent scenario, can add a level of bureaucracy to the process that results in a lack of accountability and responsibility for making sure the project is completed on time and within budget.

3. Multi-Prime Contracts

Multi-prime contract delivery systems, as the name implies, involve not one prime contract for construction of the entire project, but multiple contracts between the owner and various contractors performing separate tasks. The prime contractor in this scenario is responsible for coordinating, managing and performing a portion of the work, while the owner, or an entity designated by the owner, is responsible for the overall coordination of the project. The biggest drawback to this delivery system is coordinating the work to be performed under the separate contracts, and resolving disputes arising when one of the contractors is not performing as required, or delaying the work of others.

IV. THE BIDDING PROCESS: GETTING THE WORK

A. Overview

Owners—at least private owners—can award construction contracts any number of ways, including through direct negotiations with pre-selected contractors of their choice. One of the trademarks of the construction industry, however, is the competitive bidding process. On a competitively bid project, the owner provides a number of prospective contractors with a set of plans and specifications. Each contractor uses the plans and specifications to prepare an estimate of what it would cost to construct the project and submits a "bid" to the owner. The owner, usually with the assistance of the A/E, then reviews each bid, determines the successful bidder and awards the contract.

On private projects, the bidding process may be "closed," allowing the owner an opportunity to review each bid privately and negotiate with the various bidders before deciding which bid to accept. Most public projects, however, involve an "open" bidding process, whereby bids are opened and read publicly—usually to a room full of anxious contractors—with the contract being awarded to the "lowest responsive and responsible bidder."

The notion behind the competitive bidding process is that it allows the owner to make an "apples to apples" comparison of the prospective bidders and their bids in order to assure that the project will be constructed by a qualified contractor at the lowest price. The downside to this theory is that such competition can lead to a quasi-adversarial relationship between the owner and the contractor, where, in order to maximize its profit—or mitigate its loss—a contractor may be more inclined to assert a claim for additional costs.

In light of the continuing popularity of competitive bidding, and its predominance in the public contracting arena, the following is a brief overview of how that process generally works.

B. Bid Invitation

On a traditional design-bid-build project, the A/E prepares the "bid documents," which will later form the basis for the "contract documents." The bid documents typically include such things as the project plans and specifications, any "addenda" to the plans and specifications (i.e., corrections, clarifications, or additional information regarding the design or project conditions);[8] instructions to bidders specifying how, when, and where the bids should be prepared and submitted; the form of agreement between the owner and contractor identifying the material terms of the contract (parties, scope, price, payment terms, and time); the "general conditions" of the contract specifying the roles, responsibilities and relationships of the owner, A/E and contractor, notice requirements, payment procedures, procedures for addressing changes in the work, termination and other relevant items; the "supplementary" or "special" conditions specifying additional requirements unique to the project; geotechnical investigations and similar technical reports regarding site conditions; and payment and performance bond forms. Most of this information is usually bound together into a "project manual."

Once the bid documents have been prepared and the project is ready for bid, the owner issues a "notice to bidders" or "advertisement for bids." The advertisement is typically published in select newspapers and trade journals, often for a period of several weeks. The advertisement sets forth the type of project, its location, the project owner; when and where the bid documents are available to be picked up and their cost; the time and place for submission of bids; bid security requirements, treatment of late bids; and the time period during which bids shall remain irrevocable. The advertisement may also include requirements for

---

8. An addendum is a supplement to the bidding documents issued prior to the submission of bids and contract award. Changes to the design issued after contract award are covered by change orders.

self-performed work, requirements for use of disadvantaged or local business enterprises, contract start and completion dates and pre-bid conference or site inspection dates.

C. Bid Preparation

1. The "Form" of the Bid

Once a contractor elects to bid on a particular project, it must prepare and submit a "responsive" bid that conforms in all material respects to the owner's bidding requirements. Failure to do so will result in the bid being rejected by the owner. While the instruction-to-bidders section of the bid documents varies from project to project, typical provisions include the format in which the bid must be submitted; bid alternates, bid security requirements; payment and performance bond requirements; treatment of late bids, owner discretion to reject any and all bids; what constitutes formation of a contract; identity of subcontractors; bidder qualifications; and bidder references.

2. The "Substance" of the Bid

While compliance with the instructions to bidders places a premium on "form," the "substance" of the bid consists of the actual estimated cost to do the work. This in turn depends in part on whether the contract will be lump sum, unit price or cost plus.

a. Lump Sum Contracts

A "lump sum" contract is one in which the contractor agrees to complete a project for a set price. This is the preferred method of contracting for an owner, who knows, prior to the start of construction, the total cost of the project—barring changes in the scope of work or other unforeseen difficulties—and is in a better position to obtain financing for the project. It also places much of the risk of cost overruns on the contractor. For this reason, lump sum contracts are typically used only when the nature and scope of work is well-defined by a complete set of plans and specifications, enabling the contractor to accurately identify and quantify the materials to be installed and the tasks to be performed on the project.

b. Unit Price Contracts

A "unit price" contract typically involves a project in which the nature of the work to be performed is well-defined, but the quantities of work are not, such as in heavy or highway projects involving a significant amount of earthwork. In unit price contracts, the plans and specifications identify the various line items of work to be performed (e.g., excavation, backfill, concrete placement, etc.) and provide an engineer's estimate of the quantity of each line item of work (e.g., 1000 cubic yards of backfill). Each bidder then submits a bid that is broken down into unit price costs for performing each line item of work (e.g., backfill – $15/cubic yard). Once the project is completed, the final contract amount is determined by multiplying the total quantity of work performed for each line item by its unit price and adding up the totals. In the event that the actual quantities for a particular line item vary significantly from the engineer's estimated quantity, the contract usually permits the unit prices to be equitably adjusted.

Some contracts are a hybrid between unit price and lump sum contracts, allowing some items (e.g., foundation excavation) to be bid at a unit price, while requiring the rest of the work (e.g., construction of a multi-story office building) to be bid as a lump sum.

c. Cost Plus Contracts

A "cost plus" contract is one in which the owner agrees to reimburse the contractor for the actual costs incurred in performing the work, plus a fixed or determinable fee to cover the contractor's overhead and profit. Cost plus contracts are the antithesis of lump sum contracts, shifting much of the risk from the contractor to the owner. For this reason, they are disfavored by owners, and typically arise in situations where the owner is under time constraints to complete the project, or has an incomplete design that precludes the contractor from performing an accurate lump sum estimate. In order to temper the owner's risk on such contracts, it may insist on a guaranteed maximum price ("GMP") for the contract, possibly with a mechanism for sharing any cost savings if the contract is completed at a lower price.

d. Estimating the Contract Price

Under each of the above three contract scenarios, a contractor must account for a host of costs in estimating the price to be included in its bid. These costs include:

- Material costs for every piece of material and equipment installed on the project (e.g., plywood, concrete, valves, structural steel, doors, pumps, generators, etc.), which costs must account for sales tax and delivery;
- Labor costs for each of the various trades on the project (e.g., electricians, carpenters, pipe fitters, etc.), including direct costs associated with each trade's standard hourly wage, and indirect costs ("burden rate") associated with payroll taxes, fringe benefits, and insurance (worker's compensation, unemployment, etc.);
- Equipment costs for items ranging from small tools to tower cranes, which must take into consideration rental costs, fuel and maintenance costs, depreciation and operator costs;
- Subcontractor costs for work that will not be self-performed;
- Direct (jobsite) overhead, which includes trailers, phones/fax machines, water, electricity, office equip-

ment and supplies, project managers, superintendents, non-working foremen, secretaries, insurance, and other "non-productive" costs specific to a particular project (usually expressed in terms of a percentage);

- Indirect (home office) overhead, which includes costs of maintaining the home office (including rent, staff salaries, office supplies, equipment, storage and utilities), legal costs, advertising costs, salaries of company executives, and other similar day-to-day costs incurred in maintaining a business that cannot be charged to one particular project (usually expressed in terms of a percentage);[9]
- Markup, which is simply the amount of profit the bidder hopes to make on the project if all goes according to plan (usually expressed in terms of a percentage); and
- Bond premium costs, where bonds are required, which is customarily based upon a percentage of the overall project cost.

The relative costs of these various components will depend on the degree to which the contractor intends to self-perform the work with its own forces, or subcontract the work to other trades. On larger projects, prime contractors tend to subcontract out the bulk of the work, particularly as it pertains to the specialized trades (e.g., electricians and plumbers), and serve primarily as managers of the various trades in coordinating the performance of the work. As a result, in compiling their bids to the owner, prime contractors simply take bids from prospective subcontractors and incorporate the best subcontractor bid into their own bid.

Where the contractor elects to self-perform certain categories of work, it must perform detailed quantity take-offs

---

9. A significant amount of attention has been paid to the issue of how exactly to quantify home office overhead in terms of damages in a construction dispute, most of which centers on *Eichleay Corp.*, 60-2 B.C.A. (CCH) ¶ 2688 (1960), *aff'd on rehearing*, 61-1 B.C.A. (CCH) ¶ 2894 (1960), and its progeny. *See* Reynolds, *Is Eichleay the Answer? An In-Depth Look at Home Office Overhead Claims*, CONSTR. LAWYER 1 (April 1987); *see also* West v. All State Boiler, Inc., 146 F.3d 1368 (Fed. Cir. 1998); Melka Marine, Inc. v. United States, 187 F.3d 1370 (Fed. Cir. 1999), *cert. denied*, 529 U.S 1053 (2000).

(such cubic yards of concrete and tons of reinforcing steel) to estimate the material and labor costs of performing the work. Labor costs are calculated based upon assumed labor productivity rates in terms of how many "man-hours" it will take to perform a particular task. For example, if it takes 2 man-hours to place a cubic yard of concrete, then a project with 100,000 cubic yards of concrete would require 200,000 man-hours of labor, which, at a labor rate of $20/hour, would cost $4 million.

Once the contractor knows the total number of man-hours required, it then determines the size of the crew to employ based upon the project's time constraints. Continuing with the previous example, if a contractor has 100 weeks to complete 200,000 man-hours of concrete work, it must plan on a crew of 50 laborers, working 40 hours a week, in order to complete the work on time and without using overtime or second shifts.

Obviously, an error in judgment regarding labor productivity could prove costly, particularly where a contractor must pay overtime or hire additional laborers in order to make up for slower-than-estimated productivity rates. Likewise, projects that experience significant delay or disruption to labor productivity can quickly drain a contractor of its anticipated profit, and often result in claims.[10]

D. Bid Submission

On competitively bid projects, the owner specifies in the bid invitation the time and place for submission of bids. The bids are usually submitted in sealed envelopes along with any required bid security (bid bond), which serves as a form of guarantee on the part of the contractor that if selected, it will enter into a contract with the owner pursuant to the terms of the bid.

---

10. A number of theories have evolved and numerous articles have been written as to how to quantify and recover for this lost productivity. *See* Lynn Hawkins Patton & Cheri Turnage Gatlin, *Claims for Lost Labor Productivity*, CONSTR. LAWYER 21 (April 2000).

On public projects, the bids are opened, read aloud to the public and tabulated. Typically, however, the project is not awarded to the lowest bidder at that time. Instead, the owner and A/E take the bids and evaluate them to make sure they are "responsive" (i.e., satisfy the material requirements of the bid documents) and that the contractors submitting the bids are "responsible" (i.e., have the financial wherewithal and experience to perform the work).

During this bid evaluation period, the contractors' bids (which constitute "offers" for purposes of contract formation) remain irrevocable for a certain period of time specified in the bid documents (typically 30 to 60 days). Contractors are generally precluded from withdrawing their bids during this time period, or they run the risk of losing their bid security. The one noted exception is where a contractor's bid contains a mistake of fact (i.e., a clerical error or error in calculating/compiling figures), the mistake pertains to a material provision of the contract, the mistake is not the result of gross error on the contractor's part, and the contractor timely notifies the owner.

As noted above, the prime contractor's bid is often comprised of a number of subcontractor bids that have been incorporated into the overall bid. Under these circumstances, a prime contractor could find itself in a significant bind if, after submitting a bid to the owner, a subcontractor decides to withdraw its own bid. This is because the subcontractor's bid to the prime contractor is not irrevocable in the same manner as the prime contractor's bid to the owner is. Through the doctrine of promissory estoppel, however, courts have established an alternative means by which to hold a subcontractor accountable where a prime contractor can establish: (1) a clear and definite offer from the subcontractor (usually satisfied by the subcontractor's bid); (2) reasonable expectation on the part of the subcontractor that the contractor would rely on the subcontractor's bid; (3) actual reliance by the contractor (use of the subcontractor's bid in preparing its own bid); and (4) detriment to the contractor as a result of relying on the subcontractor's bid.[11]

Interestingly, the reverse does not hold true. The mere fact that a prime contractor incorporates a subcontractor's bid into its own bid does not obligate the prime contractor to use the subcontractor if awarded the project.[12] Nor is the subcontractor entitled to use promissory estoppel as a means of seeking equitable relief from the contractor.

V. CONTRACT FORMATION: THE CONDITIONS GOVERNING THE WORK

A. The Prime Contract

1. The Contract Documents

Unlike many commercial contracts, a construction contract is not a single document, but consists of a group of "contract documents." These contract documents typically include the agreement between the owner and contractor; the general and special conditions; the plans and specifications for the project; and any subsequent modifications (change orders) to the contract. The contract documents may also include soils reports or other engineering data. One or more of the contract documents (usually the agreement and/or the general conditions) will define what documents comprise the contract.

2. Contract Forms

---

11. The seminal case on this issue is Drennan v. Star Paving Co., 51 Cal. 2d 409, 333 P.2d 757 (Cal. 1958). However, much has been written on this issue since then. *See* Eric C. Row & Allen Holt Gwyn, *Home Electric v. Star Paving Promissory Estoppel Debate: The Rules of Offer and Acceptance Have Not Been Abandoned*, CONSTR. LAWYER 7 (Jan. 1991).

12. Although in some states, contractors are prohibited by statutes governing public procurement to substitute subcontractors that are listed in their bids to prevent "bid shopping," wherein a contractor attempts to negotiate a contract price that is lower than the original bid price from subcontractors, using the award of the contract as a leverage in urging subcontractors to lower their prices.

Larger public and private owners often draft their own contract documents. There are also a number of standard form contracts published by various industry groups, including the Associated General Contractors ("AGC"). The most frequently used contract forms, however, are those published by the American Institute of Architects ("AIA").

The AIA publishes documents for just about every relationship in construction, including standard owner/contractor agreements (AIA Document A101), cost-plus-fee owner/contractor agreements (AIA Document A111), general conditions (AIA Document A201), performance and payment bonds (AIA Document A311 and A312) and contractor/subcontractor agreements (AIA Document A401). Many of these documents are drafted to be compatible with one another. For example, AIA Document A101 (Owner/Contractor Agreement) is intended for use with AIA Document A201 (General Conditions).

The advantage in using the AIA forms from a legal perspective is that, because of their popularity, they have received a significant amount of attention in both published court opinions and legal articles as to their meaning, and parties are comfortable using them.[13] Each project is different, however, and not every party is well served by the language contained in form contracts.

Whether a standard form contract is used, or one specifically tailored for a given project, most will include terms defining/addressing the scope of work, project start and completion dates, price, payment terms, termination, change orders and extra work, delays, disputes and dispute resolution, insurance and bonding requirements, project safety, testing/inspection and correction of defective work. Within this

13. *See* Nancy N. Potter, *Design Delegation Provisions of AIA Document A201, 1997 Edition*, CONSTR. LAWYER 27 (July 1998); Patrick J. O'Conner, Jr. & William R. Joyce, *Warranties, Guarantees, and Correction Remedies under the AIA Document A201 (1997)*, CONSTR. LAWYER 19 (January 1998).

broader umbrella of contract provisions, those most frequently litigated include:

- Notice provisions, which set forth time limits within which written notices of delay and/or claims for extra work must submitted by the contractor to the owner in order to receive a time extension or an increase in the contract amount.[14] Disputes over notice provisions invariably involve issues as to what constitutes sufficient written notice, whether such notice was timely provided, and/or whether written notice requirements were waived due to the actions or inactions of the owner or A/E.
- Differing site condition clauses, allowing a contractor to recover additional costs and/or time due to concealed or subsurface conditions encountered at the site that are materially different from those represented in the contract documents ("Type I"), or of an unusual nature, materially different from those normally and customarily encountered in the type of work called for in the contract documents ("Type II").[15]
- No-damage-for-delay clauses, which limit a contractor's right to recover for project delays beyond the contractor's control to an extension of time only, with no right to recover additional costs attributable to the delay, even if the delay was the fault of the owner or the A/E.[16]
- Indemnity clauses, which obligate the contractor to indemnify the owner against claims asserted by third-

---

14. *See* AIA Document A201-1997, ¶ 4.3.2.

15. *See id.* at ¶ 4.3.4. Note also that many states have statutes addressing owner liability for differing site conditions. *See* Jeffrey M. Chu, *Differing Site Conditions: Whose Risk Are They?*, CONSTR. LAWYER 5 (April 2000).

16. While courts have upheld the validity of such exculpatory clauses, several notable exceptions to their enforceability have evolved, including: (1) delay resulting from bad faith or active interference by the owner; (2) delay beyond the contemplation of the parties; and (3) delay constituting abandonment by the owner. *See* Susan Siskind Dunne, *No Damage for Delay Clauses*, CONSTR. LAWYER 38 (April 1999); *see also* Fred D. Wilshusen & Richard Gary Thomas, *How to Beat a 'No Damage for Delay' Clause*, CONSTR. LAWYER 17 (January 1989).

parties, oftentimes even where the owner is significantly or primarily at fault.[17]

- Termination clauses, which often involve issues regarding whether a default termination was justified (i.e., who breached first?) and, more recently, issues arising from termination-for-convenience clauses and the damages recoverable.[18] (Termination clauses are addressed in detail in Chapter 10.)

B. SUBCONTRACTS

In light of the extent to which prime contractors subcontract out much of the work on the project, the terms of the subcontracts are often just as important as those of the prime contract. Like the larger public and private owners, many of the larger prime contractors have developed their own standard subcontracts. For those that have not, the AIA and the AGC offer form subcontracts.[19]

While subcontractor issues will be addressed in detail in Chapter 4 of this book, from the prime contractor's perspective, several of the more common and critical subcontract provisions regarding risk allocation include:

- Pay-if-paid (or pay-when-paid) clauses, which pre-condition the prime contractor's obligation to pay a subcontractor on the receipt of funds from the owner;[20]
- Flow-down provisions, which require the subcontractor to assume towards the prime contractor all of the

17. The validity of such indemnity clauses vary from jurisdiction to jurisdiction. Some states prohibit all such clauses as violating public policy. *See* Allen Holt Gwyn, *Anti-Indemnity Statutes*, CONSTR. LAWYER 30 (April 1992). Others permit such clauses, but draw the line where the indemnitee seeks indemnity for liability arising from its sole negligence. *See* MICH. COMP. LAWS 691.991.

18. *See* Joseph D. West, *Practical Advice Concerning the Federal Government's Termination for Convenience Clause*, CONSTR. LAWYER 26 (Oct. 1997).

19. *See* AIA Document A401.

obligations the contractor has assumed towards the owner, and *vice versa*;[21] and

- Incorporation-by-reference clauses, which expressly incorporate as part of the subcontract documents the documents comprising the prime contract.[22]

In addition to the foregoing, standard subcontract language also includes frequently litigated provisions included in prime contracts, such as arbitration clauses, liquidated damage clauses, no-damage-for-delay clauses and notice provisions.

VI. CONTRACT PERFORMANCE: DOING THE WORK

A. Staffing the Project

Once the project is awarded to a contractor, the real work begins. The principal role of the prime contractor on a construction project is management. A prime contractor that is a poor manager has little hope of finishing a project on time and within budget, or making a profit.

In order to successfully perform its management function, the contractor must adequately staff the project. Because management staff consists of "non-productive" personnel (i.e., people that do not actually place concrete, tie

---

20. These clauses have received a significant amount of attention by various state courts and legislatures. While some states have declared the clauses invalid and unenforceable through "prompt pay" legislation or judicial declaration, other state courts have upheld pay-if-paid clauses as enforceable where the subcontract makes receipt of payment from the owner an express "condition precedent" to the contractor's obligation to pay its subcontractors. Still other courts interpret the clause as requiring the contractor to pay the subcontractor within a "reasonable time," irrespective of whether the prime contractor has been paid by the owner. *See* William M. Hill & Donna M. Evans, *Pay When Paid Provisions: Still a Conundrum*, CONSTR. LAWYER 16 (April 1998); Allen Holt Gwyn, *Contractor/Subcontractor Prompt Pay Acts—Public and Private Projects*, CONSTR. LAWYER 61 (Jan. 1996).

21. Oftentimes, the terms of the prime contract mandate the inclusion of such a provision in each subcontract. *See* AIA Document A201-1997, ¶ 5.3.1.

22. *See also* T. Burt Gary, *Incorporation by Reference and Flow-Down Clauses*, CONSTR. LAWYER 1 (August 1990).

rebar, or run conduit), however, a contractor must avoid over-staffing in order to keep its overhead costs down.

Larger construction projects are usually staffed with a full-time project manager to oversee contract administration in the office and in the field and to serve as the point person in interacting with the owner, architect, and subcontractors; one or more project superintendents to oversee the actual construction work in the field; and a project engineer responsible for processing submittals and requests for information, submitting applications for payment, updating the schedule, procuring material and equipment and performing similar "office" functions. In addition, the superintendent may have several "non-working" foremen in charge of the various work areas and/or crews, while the project engineer may have one or more field engineers responsible for "laying out" the work (e.g., providing wall locations and column heights) and recording how the work is actually being installed on a set of "as-built" drawings.

B. Scheduling the Work

The most critical task in the prime contractor's management of the work is project scheduling. This process usually begins in the bidding stage, and is refined after the contract has been awarded. While the topic of scheduling is addressed in much greater detail in Chapter 8, a few comments are appropriate here given the significance of scheduling to the performance of the prime contractor.

The time frame in which the prime contractor must complete the project is set by the contract, which usually requires the project to be completed by a specific date, or within a given number of calendar days from the date the owner issues a "notice to proceed." The schedule, therefore, must identify and sequence a host of scheduling "activities" in a way that minimizes disruption and maximizes speed. In addition to the obvious need to link activities in a logical sequence (e.g., excavating foundations before placing concrete), the schedule must take into consideration the number of subcontractors that will be on site at any given time and where they will be working, in order to minimize interferences between them; accommodate and coordinate the delivery of materials to the site so they are available when needed but do not unnecessarily take up limited storage space for extended periods of time; account for seasonal weather conditions in an attempt to minimize their adverse impacts on construction activities; and coordinate the use of large pieces of equipment, such as cranes, amongst the various trades using them. Failure to perform the scheduling task competently will assure delayed project completion and claims of interference and inefficiency by subcontractors.

C. Submittals and Shop Drawings

While the plans and specifications set forth generally how the work is to be performed, the prime contractor is responsible for confirming with the A/E, through the submission of shop drawings and submittals, precisely what materials and equipment the contractor intends to use, and how they are to be fabricated and installed. Throughout the duration of the project, but particularly in the beginning, the contractor will prepare, process, and transmit submittals and shop drawings to the A/E covering nearly every material or product to be installed on the project. Each submittal, many of which originate with subcontractors and suppliers, is first reviewed by the contractor for compliance with the project design and then transmitted to the A/E for review and approval.[23] If the A/E rejects the submittal, it is returned to the contractor for revision and resubmission. Depending on

---

23. This review process can lead to disputes between the prime contractor and A/E as to who has ultimate responsibility/liability for approved shop drawings containing errors and omissions. *See* Robert A. Rubin & Mary E. Ressler, *To Build a Better Mousetrap—The Search to Define Responsibility For Shop Drawing Review*, CONSTR. LAWYER 1 (April 1985).

the nature of the submittal and whether it is approved the first time it is submitted, this review process can take weeks.

The significance of the submittal/shop drawing review process cannot be overstated. On industrial projects requiring specialized equipment, the manufacture and delivery of which may take months, the contractor can ill afford to have its order delayed by the submittal process. Therefore, the contractor must prioritize its submittals to correspond with the construction schedule and make sure they are correct the first time they are submitted.

D. Requests for Information

Submission of a Request for Information ("RFI") is the procedure through which the prime contractor can clarify errors and ambiguities in the plans and specifications. When an issue concerning the plans and specifications arises, the party impacted fills out an RFI form providing sufficient detail (e.g., specification sections and drawing numbers) to enable the A/E to identify the issue and provide clarification. If the RFI is generated by a subcontractor, the subcontractor transmits the RFI to the contractor, who reviews the RFI to confirm its validity and then transmits it to the A/E. The A/E then reviews the RFI, prepares a response, and transmits the RFI back to the contractor for distribution to those subcontractors impacted by the issue. If the A/E's response will require a change in design or change in the scope of work, the contractor will notify the A/E and owner and request a change order for additional compensation and/or time to complete the project.

The number and significance of RFIs generated on a given project is a frequent topic of litigation. Each time a contractor discovers an error or omission in the plans and specifications that must be addressed by an RFI, it must stop work and await the A/E's clarification before proceeding, a process that can take anywhere from a half-day to several weeks or months, depending upon the issue. The result for the contractor is disruption and loss of efficiency. On projects with a large number of RFIs, the "cumulative impact" of this disruption can be significant.[24]

E. Change Orders and Construction Change Directives ("CCD's")

While this topic is covered in detail in Chapter 9, a few words on change orders and construction change directives are appropriate here. By contract, the owner is allowed to add, delete and otherwise alter the work,[25] provided that such modifications are not outside the scope of the underlying contract.[26] Furthermore, RFIs and the discovery of unforeseen or differing site conditions may spawn contract modifications. Depending on the magnitude of the proposed modifications, the prime contractor may feel the need to seek an adjustment to the contract amount, contract time, or both. The procedures for implementing these adjustments to contract time and/or amount are set forth in the contract.[27] Many contracts prohibit the performance of—or at least the payment for—changed work without written authorization by the owner or its representative.

The "Change Order" is a formal, written modification to the contract, signed by the owner, the contractor, and (typically) the architect, which identifies, quantifies, and authorizes, in terms of dollars and/or time, a change in the work as originally defined by the contract documents.

---

24. *See* Geoffrey T. Keating & Thomas F. Burke, *Cumulative Impact Claims, Can They Still Succeed?*, CONSTR. LAWYER 30 (April 2000).

25. Additions to the work are known as "extras"; alterations to the work are known as "changes."

26. Changes deemed outside the scope of the contract are known as "cardinal changes" and significantly alter the rights and remedies between the owner and contractor. *See* Allied Materials & Equipment Co. v. United States, 569 F.2d 562 (Ct. Cl. 1978); Edward R. Marden Corp. v. United States, 442 F.2d 364 (Ct. Cl. 1971); C. Norman Peterson Co. v. Container Corp. of America, 218 Cal. Rptr. 592 (Cal. CT. App. 1985).

27. *See* AIA Document A201-1997, ¶ 7.

Construction Change Directives ("CCD's"), on the other hand, are written orders prepared by the A/E, and signed by both the A/E and owner, that direct the contractor to perform certain changes to the work prior to agreement by all the parties as to the adjustment in the contract amount and/or contract time. Once all parties agree to the amount/time adjustment, the CCD is converted to a Change Order.

Like RFIs, disputes regarding the existence, expense, and/or time adjustments associated with extra or changed work are a frequent subject of litigation. Examples include:

- Claims that subsurface or other conditions encountered at the site are "differing site conditions";
- Claims that the "cumulative impact" of a significant number of change orders has resulted in overall increases in the contract amount or time not sufficiently captured by the sum of the individual change orders (i.e., that the whole is greater than the sum of the parts);[28] and
- Claims arising from an owner's refusal to pay a contractor for extra work due to the contractor's failure to provide written notice/obtain written authorization.

This is but a sample. Not to be overlooked is the fact that disputes involving change orders are not limited to prime contractors, but often include subcontractors as well. Change orders often originate with a subcontractor submitting a change order request to a prime contractor, with the prime contractor in turn passing the change order upstream to the owner. Problems arise when the owner refuses to issue a change order to the prime contractor, but the subcontractor insists on an increase in its subcontract amount or time. Due to the contractual relationships between the parties, the contractor finds itself caught in the middle. One way the courts have assisted contractors in getting out of this predicament is by allowing a contractor to "sponsor" a subcontractor claim through the use of

---

28. *See* Keating & Burke, *supra* note 24, at 30.

"pass through" or "liquidating" agreements, which essentially permit the subcontractor's claim to be presented directly to the owner, while limiting—or eliminating—the contractor's exposure to the subcontractor if the claim is unsuccessful.[29]

F. Payment

If scheduling the work is a prime contractor's most important management function on the project, processing and distributing payment is not far behind. Indeed, timely processing and distribution of payment to various subcontractors and suppliers is essential in maximizing the smooth and efficient flow of work on the project. Many subcontractors and suppliers are poorly capitalized and depend upon a timely and steady stream of progress payments in order to meet payroll and procure materials. Any significant disruption to this income stream could lead to the stoppage of work, the initiation of payment bond or lien claims, or, in some cases, bankruptcy and project abandonment.

In order to minimize disruptions to the work flow, the prime contractor must make sure its monthly payment applications to the owner are submitted on a timely basis and include all the work performed through the date of the invoice. Likewise, the prime contractor must make sure that payments received from the owner are promptly dispersed, and that all subcontractors and suppliers are properly paid for their work in exchange for waivers of lien and/or bond claims. (This topic is covered in much greater detail in Chapter 7.)

---

29. The concept of pass-through or liquidating agreements originated with the decision in *Severin v. United States*, 99 Ct. Cl. 435 (1943), *cert. denied*, 322 U.S. 733 (1944), and has since received abundant attention in various court decisions and legal journals. *See* Carl A. Calvert & Carl F. Ingwalson, Jr., *Pass Through and Liquidation Agreements*, CONSTR. LAWYER 29 (October 1998).

G. Project Safety

With every news headline regarding the cave-in of a trench excavation, or the collapse of a tower crane, we are reminded of the dangers of construction. Such dangers are inherent in the nature of the work. Materials and equipment weighing thousands of pounds are hoisted high over head; iron workers assemble steel hundreds of feet in the air; saw blades whirl; rock is exploded; and extremes in weather can freeze the skin on contact or lead to heat stroke. Losses attributable to jobsite accidents are significant, both in terms of their human toll and their economic impact on the contractor in the form of increased insurance premiums, fines, lawsuits, lost productivity and low morale.[30]

While various state safety codes and federal health and safety acts, (e.g., the Occupational Safety and Health Act ("OSHA")) authorize the enforcement of safety requirements, assessment of fines, and even imprisonment for noncompliance, such action is often taken after the fact, too late to save a life or a limb. Contractors are responsible for jobsite safety, and the successful ones make it a priority day in and day out by employing project safety coordinators, implementing project safety plans, conducting weekly project safety meetings, training workers on the use of various pieces of construction equipment, keeping an ample supply of safety devices in good working condition and enforcing company safety policies.

H. Other Roles of the Contractor

In addition to the responsibilities outlined above, the prime contractor is responsible for quality control of the work that is put in place, maintaining and submitting daily and/or weekly logs and reports and conducting project meetings. At project completion, the contractor is

---

30. Marc M. Schneier, *Construction Accident Law* (1989).

also required to assemble and provide the owner with various equipment warranties and certificates, operating manuals, and a set of "as-built" drawings depicting dimensions, locations and materials as actually installed by the contractor.

VII. CONCLUSION

Far from a comprehensive discussion of all the practical or legal issues faced by prime contractors on a daily basis, the goal of this chapter is to provide an overview of the prime contractor's role on a construction project, along with a glimpse into some of the hurdles a contractor must overcome in order to construct a successful project. Construction projects provide fertile ground for claims. By gaining a practical understanding as to how these claims can originate, however, a construction lawyer is better able to aid his or her client in resolving those claims—or preventing their occurrence in the first place.

4 The Role of the Subcontractor

David A. Senter

I. INTRODUCTION

Subcontractors often feel as though they are at the bottom of the food chain in the construction world because although they usually perform most of the actual construction work, they are often perceived to have the least amount of bargaining power. They are generally smaller and less financially stable than the other parties involved in the construction process, and everything, including the money, flows down from the owner, architect, and/or general contractor before reaching the subcontractor. This section will explore the construction process from the perspective of the subcontractor.

II. ROLE OF THE SUBCONTRACTOR ON THE PROJECT

The term "subcontractor" generally defines that entity which performs labor and/or provides services that a general contractor is obligated to perform for an owner. "Specialty-trade contractors" are also typically subcontractors. In recent years, because of the role that the subcontractor plays in the project and the special skills that are required to perform technical work, certain design functions which have traditionally been the designer's function are being transferred from the design team to subcontractors. This trend towards transferring design functions to specialty-trade contractors has been the subject of many articles.[1] If the transfer is not well-conceived, fully understood, and agreed to, the process can cause numerous problems and result in partial or complete project failure.[2]

A. The "Traditional" General Contractor/Subcontractor Relationship

On a traditional construction project, an owner retains a design professional, typically an architect or an engineer, to prepare a set of construction documents consisting of "plans," which depict graphically what the contractor is to build, and "specifications" which detail the required materials and direct the general contractor's performance. Hopefully, the plans and specifications reflect the owner's objectives and will, if followed, produce the project the owner desires and for which he is paying.

Once the contract documents are prepared, the owner solicits bids from general contractors. Usually the general contractor's bid to the owner is based upon a solicitation of bids from various subcontractors to perform those portions of the specified work that the general contractor does not

1. See, e.g., Steven M. Siegfried, Esq., *Performance Specification and Contractor's Design Responsibility for Shop Drawings*, American Bar Association, Forum on the Construction Industry, Ninth Annual Meeting, "Design Issues in Construction: Who Is Responsible and What's Wrong?," San Diego, California (1993).

2. *See* Statement of the Case, Findings of Fact, Conclusions of Law, and Decision of Judge James B. Deutsch, in Missouri Bd. of Architects Professional Engineers and Land Surveyors v. Duncan, No. Ar. 84-0239 (Mo. Admin. Hearing Comm'n; Nov. 15, 1985). The "Hyatt Regency Disaster Case."

intend to perform itself. After the owner selects a general contractor (usually the lowest responsible bidder), and the contract between the owner and general contractor is executed, the general contractor enters into subcontracts with subcontractors, typically those who have submitted the lowest responsible bids in each category of work.

1. "Traditional" Roles and Responsibilities: The General Contractor

Although the general contractor contracts with the owner, the general contractor does not usually perform all of the hands-on construction work. Instead, the general contractor typically coordinates and schedules the performance of the work through subcontractors. General contractors tend to be larger and better capitalized than subcontractors. They are generally more financially stable and are perceived to have more bargaining power than subcontractors. One of the reasons that general contractors are more financially sound is because they do not generally make any advance payments for labor or materials. Instead, the terms of the construction loan and owner/general contractor contract usually contemplate periodic payments to the general contractor in the form of progress payments at stipulated intervals. These intervals are typically related in some fashion to the percentage of work completed. The general contractor, in turn, withholds its own fee, retains a percentage of the subcontractors' payments ("retainage"), and then pays the subcontractors at similar intervals for the work that they have completed.[3]

2. "Traditional" Roles and Responsibilities: The Subcontractor

Subcontractors, or specialty-trade contractors, actually perform the majority of the work on the typical project. Despite this fact, subcontractors are often not very large and do not have extensive

---

3. For a more complete discussion of the role of the general contractor on a project, see Chapter 3.

resources, leading some authorities to refer to subcontractors as the Rodney Dangerfields of the construction industry—they "don't get any respect."[4] In actuality, some subcontractors, because of the volume of their business and resources, dwarf many of the general contractors for whom they work. Nevertheless, this appears to be the exception rather than the rule, and the "typical" subcontractor can be impacted by a variety of problems, including cash flow, delay and labor shortages.

a. Cash Flow

Subcontractors tend to be smaller, more under-capitalized entities. While this is not always the case, they may not be as financially strong as either the general contractors for whom they work or the suppliers upon whom they rely for materials. Although they will not receive payments from the general contractor until specified portions of the work are completed, subcontractors usually pay their field personnel and tradesmen on a weekly basis and often must pay suppliers in advance for materials received. Placed in this financial squeeze, it is not unusual for a subcontractor to "rob Peter to pay Paul," i.e., use payments on one project to pay bills on another project.

If payment is delayed at any point in the construction chain, the subcontractor may run out of money and/or be unable to pay its tradesmen to continue working. This is why subcontractors, as a whole, file more bankruptcies than any other entity in the construction process. Thus, a project's very success or failure may depend upon whether the subcontractor is paid promptly for its work.

b. Delay

Changes in the work schedule or in the scope of work often have a severe impact on subcontractors. If a subcontractor

---

4. CONSTRUCTION SUBCONTRACTING, A LEGAL GUIDE FOR INDUSTRY PROFESSIONALS (Overton A. Currie et al. eds., John Wiley & Sons 1991).

expects that it will be able to perform work, but is delayed for a reason beyond its control, its cash flow problems will be exacerbated. Delay can cause the costs of materials to increase as a result of increased finance charges, or cause the cost of job labor to increase because workers are less efficient and take longer to complete tasks. In addition, the opportunity costs associated with delay increase overhead costs because the subcontractor will be forced to staff the current delayed job instead of another new job.

c. *Labor Shortages*

Since subcontractors are actually the front-line troops, if there is a labor shortage or shortage of qualified workers, the project schedule and quality of performance will suffer. Thus, owners and general contractors should ensure that the subcontractors hired are adequately staffed to perform the work. Similarly, in order to keep the project on schedule, they should make it a priority to pay subcontractors promptly for work completed so that the subcontractors can, in turn, compensate and retain their labor pool.

B. Alternative Subcontractor Relationships

1. Multi-Prime Contracts

Multiple prime contracting is often also referred to as parallel prime contracting. Unlike the traditional approach, in the multiple prime situation there is no general contractor. Instead, the major trade "subcontractors" (usually plumbing, electrical and mechanical) have a direct contractual relationship with the owner or the owner's agent. In this relationship, these entities are referred to as specialty-trade contractors or co-prime contractors.

Generally, multiple prime contracting is utilized by the government, its agencies and other sophisticated owners who want to eliminate the costs associated with a general contractor. Such owners must have a qualified staff to negotiate and coordinate the work that would typically otherwise be performed by the general contractor in the traditional approach. When the work on a multiple prime contract is not effectively coordinated, delays, cost overruns and other negative consequences occur,[5] because the specialty-trade contractors are unable to look to one entity (the "traditional" general contractor) who has the responsibility and authority to schedule and coordinate work, and resolve associated conflicts. Unsophisticated owners get themselves in trouble if they try to handle these tasks themselves. If it is unable to perform these tasks itself, an owner may hire one of the specialty-trade contractors or a construction manager (see below) to perform the scheduling and conflict resolution tasks to avoid these negative consequences.

Advocates of the multiple prime contract system of project delivery contend that, in general, specialty-trade contractors have the advantages of (1) being more familiar with their particular trade, the advances made and the products available, and (2) knowing how best to build a project at the least cost to the owner. Also, without the general contractor in the middle, the owner saves the general contractor's overhead and profit.

2. Construction Management Contracts

Construction management ("CM") is a relatively new method of project delivery system. "Construction management" is just what its name implies: it is a process by which the owner contracts with a CM to perform some or all of the construction management functions such as planning, scheduling, contracting, quality control and cost control. There are two basic types of construction management relationships—"agency" and "at-risk".

---

5. *See* Broadway Maintenance Corp. v. Rutgers State University, 447 A.2d 906 (N.J. 1982).

Under the "agency" CM, the owner retains most of the project risks and the CM simply performs its construction management activities as an extension of the owner's staff. An "at-risk" CM basically bears some or all of the risk of the project's success or failure.

Under the agency CM relationship, the owner would contract directly with multiple prime contractors or with a general contractor who would in turn contract with subcontractors under the traditional contracting process described above. The CM would have a separate management contract with the owner. Subcontractors would not be in privity of contract with the CM any more than they are in privity of contract with the owner. The CM would simply be managing the construction on the owner's behalf.

Under the "at-risk" CM model, the CM may contract directly with the general contractor, who in turn contracts with subcontractors. An alternative would be for the CM to contract with multiple prime contractors or specialty trade subcontractors. Under this scenario, subcontractors would be in privity of contract with the CM. In analyzing the duties and responsibilities of subcontractors under a construction management scheme, the practitioner must first understand the nature of the CM's relationship or role on the project.

III. THE BIDDING PROCESS FROM THE SUBCONTRACTOR'S PERSPECTIVE

A. In General

Under the traditional approach, the owner does not receive bids directly from subcontractors. Instead, the owner solicits a bid from a general contractor who, in turn, solicits bids from subcontractors. Just as the owner will often employ the general contractor who provides the lowest bid for the overall project, the general contractor will often employ the subcontractor who submits the lowest bid for the particular work to be subcontracted.

B. General Contractor's Use of a Subcontractor's Bid

When the general contractor relies upon the bid provided by a subcontractor to compute its own overall project bid, certain issues arise, such as whether the subcontractor's bid is irrevocable, or whether the subcontractor can withdraw its bid before the general contractor accepts it. In other words, when is the contract between the general contractor and the subcontractor formed? Even if the general contractor relies upon and includes a subcontractor's bid in its bid to the owner, older cases dealing with the issue denied relief to general contractors trying to get subcontactors to perform based on their bid, relying upon contract principles such as offer and acceptance, lack of consideration, lack certainty of terms and lack of mutuality of obligation.[6] These decisions were generally based upon the fact that even though it may have relied upon and included a subcontractor's bid in its own estimate, the general contractor did not obligate itself to use that particular subcontractor if the general contractor's bid was subsequently accepted. Thus, the general contractor had the choice of whether to go with the lowest responsible subcontractor bid, renegotiate the terms of that bid with the lowest responsible bidder, or negotiate and deal with another subcontractor. However, since the decision in *Drennan v. Star Paving Co.*,[7] courts seem more reluctant to insist on the application of contract principles and frequently use the concept of promissory estoppel to bind subcontractors to their bids where the general contractor can demonstrate reliance on the bid.

1. Reliance Resulting in Binding a Subcontractor to Its Bid (*Drennan*)

In *Drennan*, the court concluded that the use of the subcontractor's bid did not create a bilateral contract. However, the subcontractor was estopped from withdrawing its bid where the general contractor demonstrated reliance on the subcon-

---

6. *See, e.g.*, James Baird Co. v. Gimbel Bros, Inc., 64 F.2d 344 (2d Cir. 1933).
7. 333 P.2d 757 (Cal. 1958).

tractor's bid. Since *Drennan*, a slight majority of the Courts have followed the promissory estoppel principal to hold a subcontractor to its bid.[8] Under the *Drennan* rule, the general contractor can, at least for a short period of time, delay acceptance, reopen bidding with the subcontractor, and still claim the right to accept the original bid. During this period of time and until the contractor is ready to sign a contract with the subcontractor, the subcontractor has no assurance of getting the job.

a. Bid Shopping and Bid Peddling

The *Drennan* rule creates the opportunity for "bid shopping" and "bid peddling." Bid shopping allows a general contractor to reopen the process and solicit another round of lower bids from subcontractors after its bid has already been accepted by the owner. This gives the general contractor superior bargaining power over the subcontractor. Practically speaking, this rule provides the general contractor with a "protective ceiling" on the cost of work, with no obligation to use the low subcontractor's bid.

Bid peddling is the reverse of bid shopping and occurs when another subcontractor attempts to undercut the low subcontractor's bid to the general contractor. The bid peddler, instead of the general contractor, attempts to engage in a second round of bidding in order to win the work away from the low bidder. This phenomenon is sometimes also referred to in the industry as "chiseling."

b. Protection for the Subcontractor in the Bidding Process

As a result of *Drennan*, subcontractors may try to denominate their bids as a request to the general contractor for that

---

8. For a comprehensive review of all the cases dealing with this issue, see Joel Rubenstein, Esq., *Monograph on Promissory Estoppel*, American Bar Association, Forum on the Construction Industry, Eighth Annual Meeting (1992).

general contractor to make an offer. A subcontractor may also include self-serving language in its bid, such that use of the bid constitutes an acceptance that ties the general contractor to the subcontractor if the subcontractor's bid is low, or if the general contractor uses the bid in its bid to the owner. Additionally, a subcontractor can ask the general contractor for a letter of intent, which is simply a letter agreement stating that if awarded the contract, the general contractor will enter into an agreement with the subcontractor in accordance with the project "contract documents."

Finally, many states have passed legislation establishing bid depositories or other laws protecting subcontractors.[9] The rationale for such legislation derives from the fundamental principal that fairness to subcontractors will foster quality materials and workmanship, and resultant fairness to the public.

2. *Hall v. Underdown Heating & Air Cond. Co.*

Not all states follow the *Drennan* rule. For example, in North Carolina, the case of *Hall v. Underdown Heating & Air Cond. Co.*[10] establishes that promissory estoppel may not serve as a substitute for adequate consideration. In *Hall*, a general contractor had relied upon a subcontractor's bid when preparing its own bid to the owner. After the general contractor was awarded the contract, the subcontractor informed the general contractor that it was not going to do the work for which it submitted a bid. The general contractor obtained the same services from another subcontractor at a significantly higher price, and then sued the original bidder for the difference between the original bid and the amount charged by the replacement subcontractor.

---

9. *See, e.g.*, CAL. PUB. CONT. CODE § 4101 (West 1985) (protecting subcontractors' bids on public projects).

10. Home Electric Co. v. Hall and Underdown Heating and Air Cond. Co., 358 S.E.2d 539 (N.C. Ct. App. 1987), aff'd per curiam, 322 N.C. 107, 366 S.E.2d 441 (N.C. 1988).

In ruling that the original bidder could not be held to its bid absent adequate consideration, the court considered the principals of justice and the unequal bargaining power between the general contractor and the subcontractor:

> Allowing a cause of action based on promissory estoppel in construction bidding also creates the potential for injustice. It forces the subcontractor to be bound if the general contractor uses his bid, even though the general contractor is not obligated to award the job to that subcontractor. The general contractor is still free to shop around between the time he receives the subcontractor's bid and the time he needs the goods or services, to see if he can obtain them at a lower price.
>
> Using the doctrine in this context is also inequitable in that it allows the general contractor to sue the subcontractor if the subcontractor is unable to perform after the contractor has used his bid, but before he has formally accepted the subcontractor's offer. The subcontractor, however, is powerless and has no grounds on which to sue the contractor if the contractor refuses to use the subcontractor for the actual work.[11]

Since jurisdictions differ in the extent to which they will require the subcontractor to perform in accordance with its bid, the practitioner must check the applicable case law of the relevant jurisdiction before advising clients regarding the prosecution or defense of claims based on bid submissions.

IV. SUBCONTRACT AGREEMENTS

Before committing to any job, a subcontractor should review closely the terms of the agreement that will govern its performance. In far too many occasions, a subcontractor is handed a form agreement that is heavily slanted in favor of the general contractor after the job is underway, and the subcontractor has little time to consider whether the contract is fair or adequately protects its interests. This can lead to disaster if a dispute arises because the subcontractor may

---

11. *Id.* at 542, 358 S.E.2d. at 541.

be bound to a contract provision that will cost it time or money. As such, these agreements and some specific provisions warrant discussion.

A. Standard Form Agreements

1. American Institute of Architects

Over the years, several standard form agreements have been developed for use in the construction context. The most commonly encountered form agreements that have been successfully marketed in the construction industry are those prepared by the American Institute of Architects ("AIA"). These contracts are integrated documents and contemplate cross relationships among all of the project participants, even though there is no privity between the general contractor and design professional, or subcontractor and owner. The typical owner/general contractor relationship is defined in the AIA Document A201, the owner/architect relationship is reflected in the AIA Document B141, and the general contractor/subcontractor relationship is contained in the AIA Document A401. Although these form contracts reflect the typical relationships, they are easily modifiable and can be adapted to conform to almost any type of project. As a result, they provide a good starting point for most negotiations. Like most pre-printed form contracts, the AIA documents have their advantages and disadvantages.

First, the good points. Because of their widespread usage, parties in the construction industry generally feel comfortable with the AIA documents. There is no chance of a "hidden" provision. All handwritten or typewritten provisions inserted will be readily apparent. Because many special interest groups, like subcontractors, lobby the AIA when the forms are being revised, subcontractors feel, for the most part, that their concerns have been addressed and their interests have been adequately protected. Lawyers also have confidence in the standard AIA forms since many of the AIA standard provisions have been interpreted by the courts, resulting in a body of case law that can be easily referenced and reviewed. Accordingly, another advantage of these forms is that an attorney's fees for contract review and drafting will be substantially reduced when a form contract is used.

The primary disadvantage of using a specific form agreement such as the AIA Document A401 is that it may not always be appropriate to the circumstances. For instance, the A401 is more burdensome than necessary for a small residential project, since it was designed for large, commercial construction projects with multiple parties whose roles are well-defined. Furthermore, when a subcontractor has dealt with the A401 before, he may have a tendency to "gloss over" the document and may not look to see whether any additional provisions have been inserted. Finally, and most importantly, the A401 may contain clauses a subcontractor does not want to be bound by, which may result in damages unforeseen by the subcontractor.

2. Other Groups Producing Form Agreements

Other groups that publish form agreements for use between various parties to a construction project include the Engineers Joint Contract Documents Committee ("EJCDC"), the Associated General Contractors of America ("AGC"), and the Construction Owners Association of America, Inc. ("COAA"). Obviously, when analyzing any contract, especially form contracts, the practitioner is encouraged to consider the perspective of the group primarily responsible for drafting the document and advise his or her client accordingly.

B. Commonly Used/Litigated Subcontract Provisions

Many of the provisions discussed in this chapter should set off an alarm when subcontractors encounter them. Some should be stricken altogether or negotiated, but at the very least, subcontractors must be aware of the implications of these clauses.

1. Incorporation-by-Reference Clauses (Flow-up & Flow-down Provisions)

The technique of incorporation by reference is used to make provisions of one document a part of another. Typically, a subcontractor's subcontract with the general contractor will include a particular incorporation by reference provision, known as a "flow-down provision," which incorporates the terms of the general contractor's contract with the owner into the subcontract. On the other hand, a "flow-up provision" incorporates the duties owed by the owner to the general contractor into the general contractor/subcontractor agreement. As a matter of course and for protection, a subcontractor should never agree to a flow-down provision without a corresponding flow-up provision.

The following example illustrates typical flow-down and flow-up provisions:

> The Subcontractor binds itself to the Contractor for the performance of a Subcontractor's work in the same manner as the Contractor is bound to the Owner for such performance under the Contractor's contract with the Owner. The pertinent parts of such contract will be made available upon Subcontractor's request. The Contractor further assumes all duties to the Subcontractor that the Owner has assumed to the Contractor under the contract documents.

Flow-down and flow-up clauses can create certain practical problems. Subcontractors often enter into subcontracts containing flow-down provisions without ever seeing the owner-contractor agreement. When this happens, the subcontractor is agreeing to be bound by unknown provisions. The subcontractor should take all necessary steps to avoid this situation. If the general contract is for a public work, the general contract is a public record available from the governmental entity. If the job is private, the state mechanic's lien law may entitle the subcontractor to a copy of the general contract upon written demand to the owner or general contractor. Because the owner/general contractor agreement may be modified after the subcontract is signed, the subcontractor should obtain copies of the documents incorporated by reference, initial and date them, and attach them to the subcontract when it is executed.

Invariably, certain inherent ambiguities exist with respect to incorporation by reference provisions. For example, when a standard flow-down provision is used, one issue that arises is exactly what is incorporated into the subcontract from the general contract. Are only those technical specifications and drawings applicable to the subcontractor's specific work incorporated, or is the entire range of general contractor obligations, including obligations to give notice, waiver of claims, arbitration and insurance requirements incorporated as well?

When the incorporation by reference is clear and unambiguous and when reciprocal flow-up and flow-down provisions are used, the provisions will likely be enforced. If, however, ambiguity exists between specific language and incorporated language, courts may be reluctant to enforce the incorporation.[12] Consequently, incorporation by reference provisions should not serve as substitutes for specific language setting forth the rights and obligations of the parties.

2. Scope-of-Work Provisions

The scope-of-work provision sets forth the work to be done under the subcontract and all accompanying or incorporated documents, including specifications and drawings. This provision is often glossed over by subcontractors because they feel they know what is expected of them. If the subcontract is prepared by the general contractor, it will usually contain a

---

12. *See, e.g.,* Frank J. Rooney, Inc. v. Charles W. Ackerman, Inc., 219 So. 2d 110 (Fla. Dist. Ct. App. 3d Dist. 1969); Wood-Hopkins Contracting Co. v. C.H. Barco Contracting Co., 301 So. 2d 479 (Fla. Dist. Ct. App. 1st Dist. 1974); VNB Mortg. Corp. v. Lone Star Industries, Inc., 209 S.E.2d 909 (Va. 1974); *but see* MCC Powers v. Ford Motor Co., 361 S.E.2d 716 (Ga. Ct. App. 1987) (incorporating by reference a waiver of lien provision in the owner-contractor agreement).

broad scope-of-work to be performed by the subcontractor. Thus, if any disputes arise, the contractor will be able to tie the subcontractor into as much work as possible. A subcontractor should demand that a clear, concise description of the work be set forth in the subcontract. This also eliminates any overlapping of work between different subcontractors.

Because the scope-of-work often incorporates other documents, such as the plans, the specifications, and the owner/general contractor agreement, the subcontract should contain an order of priority clause. This clause ranks the various incorporated documents so that the parties will know which description governs in the case of a conflict between the documents. This will also cover problems encountered when the plans and specifications are ambiguous or erroneous. Additionally, the subcontractor should insert language stating that any work not covered by the scope-of-work clause will be subject to an "extra" or "change order," with a corresponding price adjustment and time extension.

Finally, the subcontractor should make certain that the scope-of-work covered by its bid is identical to the scope-of-work described in the subcontract. The subcontractor must insist that the language at least parallels that contained in the bid, and that any additional work be defined in specific terms. Also, the subcontractor should be sure that any items excluded from the scope-of-work contained in its bid be similarly deleted from the scope-of-work described in the subcontract.

3. Change Order and Extra Work Provisions

In almost every job, changes to the scope of a subcontractor's work will be made. These changes may occur because of concealed conditions that are found at the site, materials that may be unavailable or are no longer cost effective to use, or simply a change of mind by the owner. Hence, the subcontract should contain a provision detailing the procedure for approving change orders. On small projects, the procedure is often informal. Conversely, on larger projects, the procedure may be more complicated and may require approval by more than one party. If the contract does not specify who approves the changes, the sole power of approval generally lies with the owner.

It is important to know the distinction between a change order and an extra and to be sure that those terms, and the procedures for implementing each, are defined in the subcontract or in the general contract when incorporated by reference into the subcontract. A change order is a modification to work already contemplated by the agreement. For example, the subcontractor may propose a change in the brand of materials if the original brand is no longer available or if a comparable brand is available at significant savings. An extra, on the other hand, is an item of work beyond the original scope-of-work that is added during the course of the project.

Most construction contracts require that the subcontractor secure a written change order signed by the proper parties before proceeding with any changed or extra work. Such provisions can be valid and binding on the parties and, if not satisfied, may prevent the subcontractor from recovering compensation for changed or extra work. Thus, the subcontractor should demand a written change order, or at least a written work directive, before performing changed or extra work.

While obtaining a written change order before executing any changed work is desirable, since it protects all parties in the construction chain, it is frequently ignored in practice. The general contractor or subcontractor often cannot stop work while a formal change order is being processed. Oftentimes, realities make it necessary for the general contractor (and, in turn, the subcontractors) to rely on assurance of the owner's representative or the architect/engineer that a change order will be forthcoming.

Generally speaking, a subcontractor that performs changes or extras without the required written authorization is simply handing a recalcitrant owner or general contractor a defense to a valid claim, especially in states where such contractual provisions are strictly construed. Remember, the owner, architect and/or general contractor will always look for ways to back-charge the subcontractor. It cannot be stressed enough—no matter how small the job, it is important that change orders or extras be in writing and specifically described and that the price (or a method for determining the price) be agreed upon.

4. Notice Provisions

Other common provisions found in construction contracts are those dealing with the manner (i.e., usually in writing), timing and circumstances under which a subcontractor must provide notice to the general contractor of various events. Notice provisions are typically found throughout a construction contract and may not be readily apparent or set out in a manner which attracts one's attention. Additionally, notice requirements may vary depending upon the particular situation requiring the notice.

Notice provisions represent some of the most dangerous risk-shifting elements of a subcontract. Whether a subcontractor complies with these notice provisions, or does not, can make the difference between winning or losing a claim. It can be very easy, in the rush of activity on a construction site, for a subcontractor to overlook or ignore notice requirements. Although some courts have excused noncompliance with notice provisions in certain circumstances, many more courts have strictly construed express notice provisions to bar untimely claims. Consequently, the subcontractor must be familiar with the entire subcontract, and any incorporations in it, to ensure that it complies with all applicable notice provisions.

5. Indemnity Clauses

Most construction contracts contain a clause requiring that the general contractor indemnify the owner and its employees and agents from any claims or losses arising from the acts of the general contractor. The general contractor typically requires similar indemnifications from the subcontractors on the project. Such clauses are generally enforceable.

Some clauses, however, attempt to go further and require that the general contractor indemnify the owner from the owner's own negligence or, similarly, require that the subcontractor indemnify the general contractor from the general contractor's own negligence. Whether such clauses indemnifying a party against its own negligence are enforceable depends upon the law of the particular jurisdiction in which the project is located.[13] For example, in North Carolina, an agreement purporting to indemnify a promisee against liability arising from damages resulting, in whole or in part, from the negligence of the promisee is void as against public policy and unenforceable.[14] On the other hand, Texas has adopted the express negligence doctrine, which allows for the enforcement of such a provision if the agreement clearly expresses, without confusion and within the four corners of the document, the intent to indemnify the promisee from its own negligent acts.[15]

Regardless of the law of the jurisdiction, the subcontractor generally should attempt to strike any provision requiring it to indemnify the general contractor or owner from anything other than losses or claims arising out of the conduct of the subcontractor or its work. If such clauses are enforceable in the particular state in which the project is located, however, and the subcontractor can purchase insurance to cover the indemnity obligation, it may be a risk the subcontractor is willing to assume and for which it can be paid.

---

13. For a complete review by jurisdiction, see AMERICAN LAW FIRM ASSOCIATION, CONSTRUCTION INDUSTRY PRACTICE GROUP, A FIFTY STATE COMPENDIUM ON THE STATUTES AND CASE LAW CONCERNING INDEMNITY, STATUTES OF LIMITATION AND REPOSE, AND MECHANICS LIENS (Thomas Ferico, Esq. & Thomas Grau, Esq. eds. 1997).

14. N.C. GEN. STAT. § 22B-1.

15. *See* Ethyl Corp. v. Daniel Constr. Co., 725 S.W.2d 705 (Tex. 1987).

6. "No Damages for Delay" Clauses

The AIA owner/general contractor agreement defines "compensable delay" as delay caused by the owner which entitles the general contractor (and/or the subcontractor, if affected by the delay) not only to an extension of time, but also to compensation for the increased costs caused by the delay. Many owners try to circumvent their liability for compensable delay by including a "no damage for delay" clause in their contract with the contractor. These provisions generally state that a party to the contract waves its right to seek reimbursement for some or all forms of delay, and they are typically enforceable.[16] If the subcontract does not contain such a provision, the subcontractor should investigate whether the owner/general contractor agreement (assuming it is incorporated by reference into the subcontract) contains such a clause, or the subcontractor may unknowingly release its right to seek time and/or compensation for delay.

Sometimes, even if a contract contains what would otherwise be an enforceable "no damages for delay" clause, delay damages may be recovered if the owner unreasonably and willfully interferes with performance or misrepresents material facts, or if an unreasonably long delay occurs.[17] Additionally, some states have dealt with such provisions by statute; for example, on public projects in North Carolina, contractual language purporting to forbid or limit compensable damages for delay caused solely by the owner is unenforceable.[18]

---

16. For a detailed discussion of "no damages for delay" clauses by jurisdiction, see FIFTY STATE MONOGRAPH ON THE ENFORCEABILITY OF "NO DAMAGES FOR DELAY" CLAUSES, AMERICAN BAR ASSOCIATION, SECTION OF LITIGATION (Dennis A. Estis, Esq. and John S. Morris III, Esq. eds. 1998); *Validity And Construction of "No Damage" Clause with Respect to Delay in Building or Construction Contract*, 74 A.L.R.3d 187 (1976).

17. *See, e.g.*, Gasparini Excavating Co. v. Pennsylvania Turnpike Comm., 187 A.2d 157 (Pa. 1963); United States *ex rel.* Williams Elec. Co. v. Metric Constructors, Inc., 480 S.E.2d 447 (S.C. 1997); John E. Green Plumbing & Heating Co. v. Turner Constr. Co., 500 F. Supp. 910 (E.D. Mich. 1980), *aff'd*, 742 F.2d 965 (6th Cir. 1984).

18. See N.C. GEN. STAT. § 143-134.3; *cf.* VA. STAT. § 11-56.2.

7. Liquidated Damages Provisions

A liquidated damages provision provides for the assessment of a set sum against the subcontractor, usually on a per day basis, caused by the subcontractor's delay. The amount is determined or "liquidated" at the beginning of the job and is enforceable as long as it is not viewed as a penalty. A liquidated damages clause may be acceptable in those instances where the risk of liquidated damages may be less than the risk of actual damages. These provisions should be stricken from a subcontract, however, whenever they appear to be onerous, even if the provision includes a "bonus" if the project is completed ahead of time (although an early completion bonus certainly makes such provisions less one-sided). Not only are these provisions harmful to subcontractors, but they tend to result in poor work on the project by forcing subcontractors to accelerate work efforts to complete their jobs by specified dates. Obviously, this makes subcontractors susceptible to claims for defects and damages caused by poor workmanship.

If a subcontractor finds itself in an agreement containing a liquidated damages provision, many courts have found ways to protect them from unfair results. Although a court may hold a subcontractor liable for delay damages even though the delay was caused by other subcontractors, the owner must prove that the subcontractors' delay was a substantial factor in the delay of the project.[19] Additionally, if damages appear to be easily calculated, or if the liquidated damage amount is grossly disproportionate to the probable loss, the liquidated damages clause may be held unenforceable as a penalty.[20] The subcontractor must document delays it suffers at the hands of others, give timely notice under its

---

19. *See* Tuttle/White Constructors, Inc. v. Montgomery Elevator Co., 385 So. 2d 98 (Fla. Dist. Ct. App. 5th Dist. 1980).

20. *See, e.g.*, Knutton v. Cofield, 160 S.E.2d 29 (N.C. 1968) (enforcing liquidated damages clause after determining that said clause was not penal); *Contractual Provision for Per Diem Payments for Delay in Performance as One for Liquidated Damages or Penalty*, 12 A.L.R.4th 891 (1982).

subcontract, and make timely requests for extensions of time. This will put the subcontractor in the best possible position to defend against any delay damage claim.

8. Lien Waiver Provisions

Owners typically include waiver of lien provisions in their construction documents to ensure that their property will be not be encumbered by liens. These provisions require that the general contractor, subcontractors and/or material suppliers waive their rights to lien the project. Lien waiver agreements can be entered into either before the lien right arises (when the contract is entered into), or after the lien right arises. In most states, however, lien waivers entered into before signing the contract are unenforceable.[21]

Mechanic's lien rights are of central importance to subcontractors and, therefore, should not be waived absent payment or after having received sufficient consideration in return for the waiver. If your state does not permit waiver as consideration for entering into a contract, this is a provision you may be able to "give up" as a chip in contract negotiation since it is unenforceable anyway.

9. Insurance/Additional Insureds

Oftentimes a subcontract agreement requires that the subcontractor name the general contractor as an additional insured under its insurance program and that the subcontractor's insurance be primary. While the intent of this provision may be to provide coverage for the general contractor for vicarious liability "arising out of" the negligence of its subcontractor, the effect may be that the subcontractor's insurance is actually insuring the general contractor even for negligent acts unrelated to the subcontractor's

---

21. *See, e.g.*, FLA. STAT. § 713.20(2) (1988); N.C. GEN. STAT. § 44A-12(f), *but see* TEX PROP. CODE ANN. § 53.160(B)(7). See also *Release or Waiver of Mechanic's Lien By General Contractor as Affecting Rights of Subcontractor or Materialman*, 75 A.L.R.3d 505 (1977).

scope of work. Before agreeing to an additional insured provision, or any insurance provision, the subcontractor must review its entire insurance program to ensure its ability to provide the required coverages and that providing such contractually required coverages does not trigger an exclusion in its insurance program. Moreover, an additional insured requirement may violate the applicable state's anti-indemnity statute as discussed above. Thus, applicable state law must also be researched.[22]

10. "Pay-When-Paid" and "Pay-if-Paid" Provisions[23]

Pay-when-paid and pay-if-paid clauses are in vogue today, with the most onerous reading as follows:

> Payment by owner to contractor for subcontractor's work is a condition precedent to contractor's duty to pay subcontractor under this subcontract agreement.

When these clauses appear in a subcontract, they should cause a warning bell to go off in a subcontractor's head. Also known as contingent payment clauses, they mean that the general contractor does not have to pay the subcontractor until the owner pays the general contractor. Such clauses, as a result, unfairly shift the credit risk of nonpayment by the owner to the subcontractor, even though the subcontractor had no contract with the owner. Subcontractors should make every effort to strike such provisions from the subcontract.

In most states, these clauses must be clear and unambiguous to be enforceable against a subcontractor.[24] If the clause is ambiguous, the general contractor must pay the subcontractor within a

---

22. See Chapter 11 for a discussion of the role of the insurance on a construction project.

23. For a further discussion of this topic, see Chapter 7.

24. *See, e.g.*, Bentley Constr. Dev. & Eng'g, Inc. v. All-Phase Electric & Maintenance, Inc., 562 So. 2d 800 (Fla. Dist. Ct. App. 2nd Dist. 1990); Galloway Corp. v. S.B. Ballard Constr. Co., 464 S.E.2d 349 (V.A. 1995); Elk & Jacobs Drywall v. Town Contractors, Inc., 229 S.E.2d 260 (S.C. 1976); Koch v. Constr. Technology, Inc., 924 S.W.2d 68 (Tenn. 1996) (containing citations to similar cases in nearly every jurisdiction); FLA. STAT. § 713.245 (conditional payment bond permitted); but see Moores Bros. Co. v. Brown & Root, Inc., 207 F.3d 717 (4th Cir. 2000) (surety not entitled to assert "pay when paid" defense of principal since provision was not incorporated into bond) (applying Virginia law).

reasonable time.[25] However, even when payment is contingent, the general contractor must seek compensation from the owner in good faith and with due diligence, or the provision will be unenforceable.[26] Moreover, in some states, such clauses are unenforceable by statute as a matter of public policy.[27]

11. Retainage Provisions

Almost all construction subcontracts provide that a percentage of payments, usually 5 to 10 percent, shall be retained until some later date, such as substantial completion of the project. The purpose of retainage is to preserve a fund of money to be available to assure final completion and to correct defects in the subcontractor's work. Typically, the full percentage retained on the entire job is held until substantial completion, at which time retainage is reduced to an amount estimated to be needed to cover only the incomplete or defective items described on a written punch list.

12. Termination Provisions

Most construction contracts allow one party to terminate the agreement if the other party materially breaches the contract. The subcontract should spell out the rights, duties, and obligations of the parties in the event of contract termination. It should also define which events warrant termination. Generally speaking, nonpayment, anticipatory breach or insolvency will justify termination of the subcontract. Additionally, termination may be justified if the site was not prepared for a subcontractor[28] or if the contractor hinders a subcontractor's operations.[29]

---

25. *See Bentley*, 562 So. 2d at 801 (deeming that ninety days was a reasonable time); *Elk & Jacobs Drywall* at 413, 229 S.E.2d at 262 (leaving determination of what constitutes a "reasonable time" up to the jury).

26. *See, e.g.*, Walker v. Chancey, 117 So. 705 (Fla. 1928).

27. *See, e.g.*, N.C. GEN. STAT. § 22C-2.

28. *See, e.g.*, Great Lakes Constr. Co. v. Republic Creosoting Co., 139 F.2d 456 (8th Cir. 1943).

29. *See, e.g.*, Citizens Nat'l Bank v. Vitt, 367 F.2d 541 (5th Cir. 1966).

Sometimes an owner is given the right in a contract to terminate for "convenience." In other words, if it is economically unfeasible to continue the project, the owner may be allowed to terminate and to compensate the general contractor (and its subcontractors) for work performed and any losses incurred up to the date of termination. The amount of compensation to which the subcontractor is entitled in the event of a termination for convenience should be clearly spelled out in the subcontract agreement. The subcontractor must take particular care to ensure that it will be able clearly to document entitlement to full compensation up and through the date of the termination for convenience. Although "convenience" implies a very broad spectrum of circumstances, such a termination must be done in good faith, or the owner may have broader liability than the contract provisions contemplate.[30]

13. Arbitration Clauses

Arbitration may be designated as the sole means of resolving any disputes that arise during project construction. Again, because an arbitration provision may be incorporated by reference into the subcontract from the owner/general contractor agreement, the subcontractor is advised to review fully all project documents to gain a complete understanding of the rights, duties and obligations of all parties in the construction process.

In recent years, arbitration has become highly favored because it tends to utilize decision makers who are actively involved in and familiar with the construction industry. One of the primary arguments for arbitration of construction disputes is that many judges and juries do not understand the intricacies of the construction process. Arbitration also usually moves more quickly than the judicial system and can be much less costly than full-blown

---

30. *See* National Factors, Inc. v. United States, 492 F.2d 1383 (Ct. Cl. 1974).

litigation. However, a subcontractor with a small claim may be forced into a multiple-party arbitration that may take much longer than if its solitary claim were decided by the courts. In any event, a subcontractor should be aware of the means designated in the subcontract for dispute resolution, if any, so that it does not waste time and money filing a lawsuit in court only to find out later that the subcontract required arbitration.[31]

14. Attorneys' Fee Provisions

A subcontract may contain a provision awarding attorneys' fees to the prevailing party in any dispute under the contract. If there is an express arbitration provision in the subcontract or one that is incorporated by reference, a clause providing for attorneys' fees to the prevailing party in arbitration should also be inserted in conjunction with the arbitration clause. Otherwise, attorneys' fees incurred in arbitration may not be recoverable by the winning party.[32]

Also, the subcontractor should strike one-sided provisions that entitle the general contractor to attorneys' fees if it is the prevailing party in a lawsuit or arbitration, unless there is a reciprocal right if the subcontractor is the prevailing party. Finally, be sure to check the applicable state's mechanic's lien statutes to determine whether the party that successfully prosecutes or defends a lien action may be entitled to recover its attorneys' fees.

15. Forum Selection and Choice of Law Clauses

Sometimes, particularly when dealing with an out-of-state general contractor, a forum selection clause or a choice of law clause, or both will be inserted into the subcontract. A

31. For a further discussion of this topic, see Chapter 12.
32. *See, e.g.*, Nucor Corp. v. General Bearing Corp., 423 S.E.2d 747 (N.C. 1992).

forum selection clause determines in which courts or forum any disputes will be resolved; a choice of law clause determines which state's law will govern the contract. For example, an Illinois contractor may insert a provision that provides that all disputes will be resolved in Chicago, Illinois and that the law of the state of Illinois is to govern the contract, even though the project is in some other state. Thus, there is at least the potential that a subcontractor could subject itself to a state with laws unfavorable to subcontractors under a particular set of circumstances. This is yet another reason why it is important for a subcontractor to review the owner/general contractor agreement if it is incorporated by reference into the subcontract.

Furthermore, before assuming that such forum selection or choice of law clauses are valid, the subcontractor should investigate the laws of the state in which the project is located and/or the contract was executed. Some states have statutes prohibiting the use of certain types of these selection clauses.[33]

V. CONTRACT PERFORMANCE

A. Changes

As noted above, projects are rarely, if ever, built exactly as shown in the plans and specifications. Changes are almost inevitable. Discussed below are some of the mechanisms for dealing with these changes.

---

33. *See, e.g.*, N.C. GEN. STAT. § 22B-3; VA. CODE ANN. § 8.01-262.1(B); *see also Validity and Effect of Stipulation in Contract to Effect That It Shall Be Governed by Law of Particular State Which Is Neither Place Where Contract Is Made nor Place Where It Is to Be Performed*, 16 A.L.R.4th 967 (1982); *Validity and Effect of Stipulation in Contract Provision Limiting Place or Court in Which Action May Be Brought*, 31 A.L.R.4th 404 (1984); *but see* Stewart Org., Inc. v. Ricoh Corp., 487 U.S. 22 (1988) (state policy disfavoring forum selection clauses does not invalidate such clauses but is merely one factor in the multi-factor analysis conducted under 28 U.S.C. § 1404).

1. Field Changes

Provisions pertaining to field changes and change orders must be read in conjunction with notice provisions in order to protect the subcontractor's rights to payments. Field changes, meaning changes in work ordered by the owner, architect or contractor without formal change orders, are often referred to as "modifications" or "minor changes." For example, paragraph 5.1 of AIA Document A401 (1997 ed.) provides:

> 5.1. The Owner may make changes in the Work by issuing Modifications to the Prime Contract. Upon receipt of such a Modification issued subsequent to the execution of the Subcontract Agreement, the Contractor shall promptly notify the Subcontractor of the Modification. Unless otherwise directed by the Contractor, the Subcontractor shall not thereafter order materials or perform Work which would be inconsistent with the changes made by the Modifications to the Prime Contract.

2. Change Orders and Constructive Change Directives ("CCD's")

Modifications or field changes are often used by the owner or architect where they believe the change will not involve extra work or delay. Change orders, on the other hand, are used where the change involves an increase or deduction in contract price or time. There is often disagreement among the owner, architect, contractor and subcontractors as to whether a modification should involve a change order. If such a dispute arises, it is critically important for the claimant to give notice of its claim and to perfect its request for a change order in accordance with contract requirements.

The AIA A201 General Conditions document recognizes this common dispute, and provides still another device, called a construction change directive ("CCD"), to effect a change.[34] The CCD allows the owner and architect to order a change without any presumption regarding whether it requires an

---

34. *See* AIA Document A201 ¶ 7.3 (1997).

adjustment in contract price or time. The device spells out the general contractor's steps to make a claim for extras, or to accept the adjustment, if any, offered by the owner. In any event, the general contractor must act quickly upon issuance of a CCD and not confuse it with a change order. If the general contractor does not promptly make its claim for an extra, it may waive its right to prosecute and prove its claim, which will be decided by the architect.

CCD's will typically affect a subcontractor more adversely than the general contractor, because the subcontractor is performing the work. For subcontractors, this provision in the 1997 edition of the AIA form documents makes it easier for an owner to request a change and to compel prompt performance without first resolving the cost and time impact of the change. Accordingly, the CCD provision creates an additional notice requirement, and therefore still another way that the subcontractor may lose its right to be paid.[35]

B. Acceleration, Out of Sequence and Delay Costs

1. Labor Estimates: Background

One of the most elusive ways for a subcontractor to lose money on a job is from increased labor and overhead. Labor is the most difficult aspect of a job to estimate, and the causes of increased labor costs are among the most difficult to trace. Faced with a subcontractor's claim for increased labor costs on a job, owners and general contractors routinely answer that the subcontractor underestimated its labor for the job, or that the subcontractor's forces were inefficient because of inexperience, insufficient equipment, failure of the sub to supervise its forces or to man the job adequately. Further, all jobs experience some delays, disruptions and acceleration, not all of which are compensable.

---

35. For further discussion on this topic, see Chapter 9.

Assuming the subcontractor did not underbid its labor and that its crews were reasonably efficient, the subcontractor will often point to the contractor's poor scheduling as the cause of the delayed, accelerated, disrupted, out of sequence or stacked work. Scheduling problems are among the most costly on a large construction project, and care must be taken to identify the relevant subcontract provisions which govern them. No contractual provision which refers to scheduling should be ignored, for any such provision may provide the basis for a claim or a defense to a claim.

Increased costs attendant to scheduling problems are best analyzed as extras. It is important, however, to segregate the types of damages arising from cost-increasing events, primarily because of subcontract provisions limiting recovery of certain costs. For example, a subcontractor faced with a binding "no damage for delay" clause may still recover much of his increased costs if he can successfully typecast them as acceleration or disruption damages, rather than delay damages. Likewise, a subcontractor faced with an onerous scheduling clause, allowing the general contractor to dictate overtime, shift work, stack trades and the like, might base its claim for increased costs on changed conditions or incomplete and inadequate plans and specifications.

The subcontractor who plots out its performance and payment plan for the job will know immediately when its costs are exceeding those planned by comparing its expected costs against the actual completion and payment percentages at any given time during the job. If it sees that its costs are greater than the completion and payment percentages, it should immediately determine if it has a claim and then initiate its claim notice under the subcontract claim procedure. Just as importantly, it should ask for an extension of time, if warranted.

2. Scheduling Obligation

Ordinarily, the general contractor is obligated to schedule the work without hindering or harming subcontractors. This obligation may be expressly imposed by contract[36] or implied by law. Some courts express the implied obligation in terms of the general contractor's obligation to make good all damages suffered by subcontractors as a result of delays caused by the general contractor. A greater number of courts speak of the general contractor's duty to refrain from doing any act that would hinder or prevent the subcontractor's performance of the work.

In general, any express provision in the subcontract concerning coordination or expedition of the work is given greater weight than the implied obligation not to delay the work.[37] Therefore, where the general contractor's right to schedule is expressed in less favorable terms than the AIA form contracts referenced above, a subcontractor's right to recover increased costs may be barred.

3. Limitations to Duty Not to Delay

Even where the subcontract does not contain express scheduling provisions or a "no damage for delay" clause, the subcontractor may still face defenses to a delay claim. The general contractor's obligation not to delay the work of the subcontractors is limited in at least two ways. First, absent an express contractual commitment to the contrary, a general contractor is not liable to subcontractors for delay damages unless the delays are caused by the general contractor or some agent or circumstance under his control.

Second, the general contractor may not be liable for delays caused by the owner or by the agents of the owner such as the architect or engineer. What are the subcontractor's remedies? They vary by state. In North Carolina, for example, the subcontractor may have a direct action against the injuring architect. In Virginia, however, this action is barred if the delay damages are viewed as "mere" economic harm.

---

36. *See* AIA Document A201 ¶ 3.3.1 (1997); AIA Document A401 ¶ 3.3.1 (1997).

37. *See Building and Construction Contracts: Prime Contractor's Liability to Subcontractor for Delay in Performance*, 16 A.L.R.3d 1252 (1967).

Even if the delays suffered by a subcontractor are due to actions of the general contractor or circumstances under the general contractor's control, the general contractor may not be liable if the general contractor's conduct evidenced a good faith effort and reasonable diligence to avoid delays. The implied obligation not to hinder the work does not impose a strict or absolute standard of conduct on the general contractor. Courts recognize that delays are not unusual on construction projects and that reasonable subcontractors anticipate some delay as a risk of doing business. If the contract expressly provides that the general contractor undertakes a specific obligation to expedite the work, however, the general contractor may be liable for damages caused by delays resulting from the failure to fulfill the express obligations, regardless of good faith efforts or reasonable diligence.

C. Shop Drawings and Submittal Requirements

1. Shifted Design Burden

Shop drawing and submittal requirements which attempt to shift responsibility and the risk for problems caused by them to the subcontractors are also on the rise. In an attempt to reduce their design costs and speed up construction, owners today are reducing the pre-bid scope of work of their design professionals, and what results is sometimes a correspondingly less complete design. Drawings are becoming more schematic, and specifications relate more to performance and "intent" than to specific items required to be included in the structure. The necessity of fleshing out the design professional's intent to build a completed project is therefore passed on to the general contractor, who then shifts the responsibility to the subcontractors by way of dragnet clauses and stringent submittal requirements. The problem is magnified by the likely lack of insurance available to a subcontractor if it engages in design work since the subcontractor's standard liability insurance policy will most likely exclude coverage for "design services."

In such situations, the design is not finalized until the design professional approves the subcontractor's shop drawings, and it is left to the subcontractor to make certain his planned equipment and method of construction meet the design professional's requirements. Where the design professional's duties of shop drawing review are also strictly budgeted, the subcontractor's ability to get needed cooperation is also limited. Fast track projects are replete with these occurrences, because the primary aim is to get the project started as soon as possible, and often on a strict budget.

2. Dragnet Clauses

Before accepting a subcontract based on incomplete plans and specifications, a subcontractor must be wary of dragnet clauses which impact its scope of work. Dragnet clauses requiring the subcontractor to perform work not expressly detailed in the plans and specifications may contain much more work than its estimate. For example: (a) . . . all mechanical work, (b) . . . and other related items, and (c) . . . all work incidental thereto.

As in scheduling, the economic risk to a subcontractor for shop drawing and submittal requirements is often not recognized until problems develop on the job. The subcontractor takes the risk that its interpretation of the design is consistent with that of the design professional, that substitutes will be allowed, and that the design professional will quickly turn around the submittal information so as not to delay the project.

Meanwhile, shop drawing delay, including disapproval of submittals requiring resubmittals and/or substitutes, puts additional risk on the subcontractor that the general contractor or other subcontractors will claim delay and increased costs, which, through other subcontract provisions, the general contractor may use to withhold payment from or backcharge the subcontractor. Even in situations where the subcontractor is clearly due an extra, its only recourse may be to pass through a claim against the owner.

3. Performance and Submittal Requirements

Subcontract provisions relating to shop drawings mirror the industry-wide shift of responsibility for a project's systems to the furnishing and installing subcontractor, for even after approval, the work must conform to the performance requirements of the Contract Documents. Shop drawing requirements differ between subcontractors, with structural steel, mechanical and electrical subcontractors being most directly affected.

To the extent that the subcontract also includes contingent payment clauses, however, even a subcontractor with virtually no submittal requirements can have its progress and final payments affected by the failure of another subcontractor to perform the submittal requirements efficiently. Moreover, these provisions may be used by the general contractor as a shield against acceleration or delay claims to the extent delay in approving the submittals is concurrent with compensable delay arising from other causes.

D. Site Inspection, Acceptance of Underlying Work and Site Conditions

1. Site Inspection/Condition

Nearly every construction contract will contain provisions requiring the contractor to inspect the site, the contract documents and the preceding work of others on which its work will be placed. Many such provisions go beyond requiring an inspection, and state conclusively that such inspection has taken place. Such provisions are often placed in subcontracts, or are incorporated from the general contract as discussed above.[38]

The subcontractor can best avoid the risk of unsuitable site or underlying work conditions by meticulously inspecting the site or the work. This is often impracticable, however, due to limitations of time, or because no meaningful inspection can

---

38. *See* AIA Document A201 ¶ 4.3.4 (1997).

be made of all underlying conditions. Such inspection by the subcontractor also may be impossible where the subject of inspection is covered up, or where the subcontractor is expected to have accepted the site inspection performed by the general contractor, through operation of incorporated provisions. In impracticable or impossible situations, the hardship on the subcontractor of these conditions must be fully developed in making the claim, and the exceptions provided by case law utilized to the fullest extent to protect and enforce the subcontractor's right to an extra.

The purpose of these provisions is to place on the subcontractor the risk of unsuitable site conditions or underlying work and the risk that its estimate includes all items relevant to its scope of work, wherever found in the plans and specifications. The subcontractor should try to protect itself against these risks in the beginning by insisting upon an acceptable "differing site condition" clause in the subcontract, such as is found in section 4.3.4 of AIA A201 (1997).

2. Acceptance of Underlying Work

By inspecting the site or underlying work, and then proceeding to place its own work, the subcontractor will likely be deemed to have accepted the site or underlying work, along with the responsibility to correct its work if the underlying site or work proves unsuitable. By accepting one of these provisions, together with a dragnet clause such as "all mechanical work shown," the subcontractor runs the risk that the plans or specifications include portions of its work in places where the subcontractor would not expect to find them.

Unfortunately, architects frequently place work on the drawings for a different category of drawings, especially where trades overlap, such as putting mechanical items on electrical drawings and vice versa. Subcontractors can best avoid the risk of enlargement of their scope of work in the bid by precisely describing the work involved, for instance, by limiting the work to that shown on certain drawings, and not others.

Acceptance of underlying work clauses is of particular importance to subcontractors such as paving, floor tile or painting trades, who must install their work over the work of grading, concrete or drywall trades. For example, if the floor tile subcontractor has entered into a contract providing (expressly or by incorporation) that it has fully inspected the work (e.g., concrete floors) upon which its work will be placed, its claim for extra compensation for additional work to make the tile fit because of defective concrete floors may be denied; or worse, its work may be rejected even though the defect lies in the concrete.

VI. DISPUTES AND REMEDIES

Federal and state statutes provide protection to subcontractors by requiring bonds in certain circumstances and by allowing subcontractors to impose liens on funds and/or real property in certain circumstances. Subcontractors may also negotiate for additional protection, such as a joint check arrangement.

A. Mechanic's Liens

A mechanic's lien is an effective method to protect suppliers of labor or material for a private construction project against non-payment for their goods or services. Generally speaking, a mechanic's lien is a statutory claim or charge on real property for the payment of a debt due to a contractor, subcontractor, material supplier, design professional or other party who supplies labor or materials to improve the real property.

Mechanic's liens are purely statutory devices and are available in every state. Most mechanic's lien statutes differ with regard to whom is protected, the property interests that can be liened, and the methods by which a lienor can establish and enforce its claim. Because liens are creatures of statute, most states rely upon the doctrine of strict construction to interpret mechanic's lien statutes.[39] Once a determination has been made that a lien has attached, most courts will apply a liberal construction of the statute to accomplish the statute's objectives.[40] Some states' statutes, however, mandate strict interpretation nevertheless.[41] Because there is significant variation between statutes in the various states, the practitioner is cautioned to examine the applicable mechanic's lien statutes (and case law interpretations of them) or other relevant statutes for specific application.

1. Classes of Liens Generally

Mechanic's lien laws are generally divided into two classes, one known as the "New York system," and the other as the "Pennsylvania system."[42] The primary distinction between the two systems is this: under the "New York system," the subcontractor cannot recover more than is due from the owner to the general contractor, while under the "Pennsylvania system," the original contract, or payment to the original contractor is no defense to a claim of a subcontractor. Under the Pennsylvania system, the subcontractor and the materialman each have a direct lien, while under the New York system, no one except the general contractor has an absolute lien, but a species of equitable subrogation is allowed.

2. Definitions

The first step in analyzing a particular mechanic's lien statute is to review the "definitions" section. Entitlement to a mechanic's lien often varies by state depending upon the statutory definition

---

39. *See, e.g.*, Schmidt v. Anderson, 97 N.E. 291 (Ill. 1911); Balco Corp. v. D.H. Overmyer Co., 334 N.E.2d 484 (Ohio Ct. App. 1975); Cameron & Barkley Co. v. American Ins. Co., 434 S.E.2d 632 (N.C. Ct. App.)1993).

40. *See, e.g.*, Talco Capital Corp. v. State Underground Parking Comm., 41 Ohio App. 2d 171, 324 N.E.2d 762 (1974); Carolina Builders Corp. v. Howard-Veasey Homes, Inc., 72 N.C. App. 224, 324 S.E.2d 626, (1985).

41. *See* FLA. STAT. § 713.37 (disfavoring liberal construction of lien statute).

42. *See Direct or Derivative Nature of Lien of Subcontractor, Materialman, or Laborer*, 53 AM. JUR. 2D *Mechanics' Liens* § 8 (1996).

of the lienor and upon the type of improvement or interest of the "owner" in the improvement.

The practitioner is wise to take particular care with respect to reviewing the definitions of subcontractors and material suppliers/materialmen and the tier in which they fall in the construction chain. Oftentimes how the tiered subcontractors/suppliers are defined determines whether they are afforded protection by the mechanic's lien statutes. For example, in Pennsylvania, a "subcontractor" is defined as one who contracts directly with the general contractor (i.e., a first-tier subcontractor). Moreover, the definition expressly excludes "a person who contracts with a subcontractor or with a materialman."[43] A further provision of the Pennsylvania lien statutes states that "[n]o lien shall be allowed in favor of any person other than a contractor or subcontractor, as defined herein, even though such person furnishes labor or materials to an improvement."[44] Thus, in Pennsylvania, lien rights are extended only to first-tier subcontractors/suppliers.

In states where the statutes do not expressly address lower tier subcontractors/suppliers, the courts often take the position that a material supplier who supplies another material supplier (as opposed to a subcontractor working on the project) does not directly contribute to a project, and therefore is not entitled to claim a lien.[45] Thus, it is important to review not only the statutory definitions, but also the case law interpreting them.

3. Contract Requirement

Under most states' lien statutes, a lien can attach against an estate in land only if the improvement was made with express or implied consent of the owner or its agent under a valid and enforceable contract. Whether the "contract"

---

43. 49 PA. STAT. § 1201(5).
44. 49 PA. STAT. § 1303(a).
45. *See, e.g.*, Evansville v. Verplank Concrete & Supply, Inc., 400 N.E.2d 812 (Ind. Ct. App. 1980); *but see* N.C. GEN. STAT. §§ 44A-8 and 44A-18 (providing lower tiered subcontractors with lien rights).

requirement has been satisfied can be resolved by application of general principles of contract law. While a claimant's "contract" need not be in writing, it is advisable to have a written purchase order or invoice which specifically designates the real property for which the lien claimant's labor and materials are being provided. This serves to provide an indication that a lien claimant is relying on that particular real property as security for payment.

4. Property Interests

Generally speaking, a lien will not attach without the consent of the owner of the property being improved. The theory is that the property should not be encumbered by the lien unless the improvements were made, or the labor or materials supplied at the direction of or with the consent of the owner of the property. Sometimes, it is not always apparent whether the party contracting for or directing the improvements is actually the owner of the property being improved. For a subcontractor, who is at least one step removed (by the general contractor) from the party directing the work, it may be even less clear. Nevertheless, the subcontractor or supplier should make every effort to determine exactly who owns the property being improved prior to commencing work.

One issue that frequently arises in this context is whether a claimant who deals with the lessee of real property and otherwise complies with the lien statute has any lien rights against the real property improved.[46] Some states address this issue by statute. For example, Florida's lien law exempts the owner's/lessor's interest from the lien of any lienor providing labor, services or materials to a lessee unless the owner/lessor joins in the contract for the improvement.[47] Under that same section, an owner/lessor would be required to record a Notice to

---

46. For a detailed discussion of the issue, see *Enforceability of Mechanic's Lien Attached to Leasehold Estate Against Landlord's Fee*, 74 A.L.R.3d 330 (1976).

47. *See* FLA. STAT. § 713.10.

Lienors reflecting that its lease provides for the exemption in order for the exemption to be operative.

In states in which it is not expressly addressed by statute, the issue is often determined by the courts. For example, the North Carolina Supreme Court has ruled that the relationship of lessor and lessee is not that of principal and agent sufficient to encumber a fee simple interest in the property improved at the request of the lessee.[48] This is true even though the owner/lessor had knowledge of the improvements, unless the claimant produces evidence that the owner/lessor had reason to believe that the claimant would look to the owner/lessor for payment.[49]

5. Improvements

Not all improvements are lienable. Determining the types of improvements subject to liens necessarily requires a review of the particular jurisdiction's lien statutes. In general, most states either require that materials be permanently affixed to the real property or be an integral part of, or incorporated into, the structure or improvement. Some categories of particular interest are discussed below.

a. Materials Delivered to Site or Off-Site

Most statutes provide that proof of delivery of materials to a jobsite constitutes prima facie evidence of actual incorporation into the improvement, thereby providing a basis for a lien.[50] The lien can usually still be defeated by proof that the materials, however, although delivered, were not incorporated into the improvement. While materials delivered to an off-site location may not generally form the basis for a lien, some statutes may create rebuttable

48. *See* Brown v. Ward, 20 S.E.2d 324 (N.C. 1942).

49. *See id.* at 346–47, 20 S.E.2d at 326; *see also* Dowd H. Price Fixture Co. v. Flowers & Monroe, Inc., 263, 175 S.E.2d 695, 696 (N.C. App. 1970).

50. *See, e.g.*, FLA. STAT. § 713.08(1)(c); *see also* Contract Steel Sales, Inc. v. Freedom Constr. Co., 362 S.E.2d 547 (N.C. 1987)

presumptions of incorporation where the owner directs that materials be delivered and stored at a specific off-site location.

b. Specially Fabricated Materials

Specially fabricated materials are those materials that are not only specifically ordered, but which are fabricated for a particular job, and have no use other than for the specific improvement. Many statutes recognize a lien for specially-fabricated materials even though they were never incorporated into the improvement or delivered at the jobsite.[51]

c. Rental Equipment

Some statutes permit a lien for the reasonable rental value of equipment used in construction, while others require that there be some labor furnished in conjunction with the use of the rental equipment.

d. Overhead, Profit and Delay Damages

Most statutes only provide a lien for labor, services and materials which improve the property. Thus, a subcontractor's usually may not include unpaid overhead, lost profit or home office overhead damages in its lien. Where these costs are important to the subcontractor's claim, however, the practitioner should review the relevant statutes and case law to see if they can be included in the lien amounts.

6. Notice of Commencement/Furnishing

A preliminary "Notice of Commencement," "Notice to Owner" or "Notice of Intent to Furnish Services or Materials," is often a prerequisite to the filing and enforcement

51. *See, e.g.*, FLA. STAT. § 713.08(1)(c); TEX. PROP. CODE ANN. § 53.021(a)(1).

of the Claim of Lien. Generally, this pre-lien notice is only required of claimants who are not in privity of contract with the owner. The purpose of the notice is to alert the owner of all potential claimants on the job that may be filing liens if they are not compensated for services rendered or materials supplied. General contractors, professional lienors and other exempt groups (i.e., laborers), may not be required to provide this notice.

The preliminary notice serves to inform the owner of the name and address of the claimant, the nature and amount of the services and materials to be furnished, and such other information as may be required by the particular statute. Most statutes require the notice to be in writing and mandate the manner and time limits for service of the notice on the owner. For example, in Florida, in most situations, the Notice to Owner must be served upon all entities above the party with whom the claimant contracted by registered or certified mail within 45 days after the claimant commences to furnish labor or materials.[52] In California, the period is 20 days.

The failure to post and/or serve such a notice timely will often prevent recovery. Therefore, a close analysis of the particular state's procedural requirements relating to the perfection of the lien cannot be over emphasized.

7. Foreclosure Action

The procedures for actually enforcing a lien are also state specific. In most states, suit to foreclose the lien claim must be filed within the statutorily prescribed time period, or the lien is expunged. This suit deadline is usually measured as a period of time (e.g., 180 days) following the date of the claimant's last furnishing of labor or materials.

52. *See* FLA. STAT. § 713.06(2)(a).

8. Attorneys' Fees

Consult the applicable state statutes to determine whether attorneys' fees may be collected in the prosecution or defense of lien enforcement actions. Many statutes grant courts discretion to award attorneys' fees to the prevailing party in certain circumstances (e.g., if there has been an "unreasonable refusal by the losing party to fully resolve the matter which constituted the basis of the suit,"[53] or if it would be "equitable and just."[54]).

B. Stop Notices/Liens on Funds

In addition to a lien that attaches to the real property being improved, some states have adopted so-called "stop notice" statutes.[55] These statutory provisions allow the unpaid party, typically a subcontractor or materialman, to stop, "freeze" or "trap" construction funds before they are disbursed by giving a notice of claim to the person holding these funds, typically the owner or the lender who is financing the project. Some statutes impose personal liability on the fund holder if it disperses funds in spite of and after having received the "stop notice." Strict compliance with timing and service requirements is required in order to enforce a "stop notice," so examination of the relevant jurisdiction's statutes and case law is a must.

C. An Alternative Remedy—Payment Bonds

1. The Federal Miller Act and "Little Miller Acts"

Publicly owned property is generally not subject to mechanic's liens. Protection is afforded those who furnish labor and materials for public construction projects either

---

53. N.C. GEN. STAT. § 44A-35.

54. TEX. PROP. CODE ANN. § 53.156.

55. For a detailed discussion of the issue by jurisdiction, see *Construction and Effect of Statutes Requiring Construction Fundholder to Withhold Payments upon "Stop Notice" from Subcontractor, Materialman, or Other Person Entitled to Funds*, 4 A.L.R.5th 772 (1993). *See also, e.g.*, N.C. GEN. STAT. §§ 44A-18, 19; TEX. PROP. CODE ANN. § 53.081.

through statutorily authorized stop notices, or through the application of the Federal Miller Act[56] on federal projects, or through state "Little Miller Acts" on state-owned projects. Not only do state "Little Miller Acts" substantially mirror the Federal Miller Act, but most state courts will turn to federal cases for guidance when questions arise regarding application of the state statutes. Because of this similarity of terms and application, this discussion will focus primarily on the Federal Miller Act. The practitioner, however, is cautioned to review a particular state's Little Miller Act for specific prerequisites for making a payment bond claim in that state.

2. Types of Bonds Required

The Federal Miller Act requires the general contractor on a federal construction project to furnish two types of bonds: a performance bond running to the benefit of the United States and ensuring the performance of the contract, and a payment bond guaranteeing payment to subcontractors and suppliers for labor or materials furnished if the general contractor defaults. With few exceptions, the furnishing of bonds is a non-waivable requirement.

3. Amount of Payment Bond

The Federal Miller Act provides that the amount of the payment bond furnished by the general contractor will be equal to "the total amount payable by the terms of the contract unless the contracting officer awarding the contract makes a written determination supported by specific findings that a payment bond in that amount is impractical, in which case the amount of the payment bond shall be set by the contracting officer."[57]

---

56. 40 U.S.C. § 270a *et seq.* (1999).
57. 40 U.S.C. § 270a(a)(2).

4. Coverage of Payment Bond

Although one section of the Federal Miller Act requires a payment bond "for the protection of all persons supplying labor and material in the prosecution of the work,"[58] another provides that the coverage only extends to those supplying labor and materials to the general contractor or to a first-tier subcontractor.[59] Consequently, only the following parties are covered by the payment bond:

- material suppliers who contract with the general contractor;
- first-tier subcontractors who contract with the general contractor; and
- second-tier subcontractors and material suppliers who contract with first-tier subcontractors.

Note that material suppliers who contract with material suppliers are *not* covered by a Federal Miller Act payment bond.

Definitive proof of the actual incorporation or consumption of materials is not required.[60] The claimant may recover for all materials supplied when there is a good faith belief that the materials were intended for use in the public project.[61]

5. Notice Requirements

The Miller Act provides that a claimant not in privity with the general contractor must give written notice by registered mail to the general contractor of any non-payment, including the amount claimed and the name of the party to whom the materials were supplied, within ninety (90) days from the date on which the last labor and materials were furnished by

---

58. Id.

59. *See* 40 U.S.C. § 270a(a)(2); *see also* Clifford F. MacEvoy Co. v. United States, 322 U.S. 102 (1944).

60. *See, e.g.*, United States *ex rel.* Lanahan Lumber Co. v. Spearin, Preston & Burrows, Inc., 496 F. Supp. 816 (M.D. Fla. 1980); United States *ex rel.* McDermott, Inc. v. Woods Constr. Co., 224 F. Supp. 406 (N.D. Okla. 1963); United States *ex rel.* Westinghouse Electric Supply Co. v. Endebrock-White Co., 275 F.2d 57 (4th Cir. 1960).

61. *See, e.g.*, United States *ex rel.* Balzer Pacific Equipment Co. v. Fidelity & Deposit Co., 895 F.2d 546 (9th Cir. 1990).

the claimant under the original contract.[62] A claimant must institute legal action against the Miller Act payment bond within one year from the date the last labor was performed or last materials were supplied. Failure to comply strictly with statutory deadlines will result in a loss of bond rights.

There is a split of authority among the Federal circuits as to whether this notice must contain a demand for payment.[63] Prudent counsel will include such a demand in its notice letter.

6. Jurisdiction and Venue

Jurisdiction and venue for a Miller Act payment bond action lies exclusively in the United States District Court "for any district in which the contract was to be performed and executed and not elsewhere."[64]

D. Payment Bonds on Non-Public Projects

Many mechanic's lien statutes provide that an owner, in order to exempt its property from the lien of all lienors other than the general contractor, may have the general contractor furnish and record a statutory payment bond. This payment bond serves as a substitute for a lien against the realty and, in some instances, may be a complete defense to a mechanic's lien action.

Additionally, although not statutorily required, general contractors and subcontractors on many private construction projects often furnish payment bonds to protect all those who contract with or under them. These non-statutory bonds are sometimes referred to as common-law bonds. Claimants under statutory bonds are required to comply with the notice requirements contained in the statute, while claimants under non-statutory bonds are required to comply with the conditions and terms contained in the bond.

---

62. *See*, 40 U.S.C. § 270b(a).

63. *See, e.g.*, United States *ex rel.* S&G Excavating, Inc. v. Seaboard Sur. Co., 236 F.3d 883 (7th Cir. 2001).

64. 40 U.S.C. § 270b(b).

E. Pass-Through Claims Against the Owner

1. Liquidating Agreements

Another cost-effective device for subcontractors is a liquidating agreement. These agreements are used when the subcontractor is trying to obtain payment from its general contractor, and the general contractor considers itself an innocent party involved in a larger dispute involving the owner, architect/engineer or other third party. In this situation, it is advantageous for the general contractor and subcontractors to band together in the litigation. The parties can enter into a liquidating agreement wherein they agree to split the legal bills and the subcontractor agrees to absolve the general contractor from liability unless the general contractor receives payment from the owner. In many of these agreements, the subcontractors assign their lien rights to the general contractor to pursue in the litigation. Liquidating agreements provide a cost-effective means of litigating a large dispute and create a united front to battle the defaulting party.[65]

2. The *Severin* Doctrine

On Federal projects, subcontractors do not have standing to sue the government because they are not in privity of contract with the contracting body—only the general contractor is. Oftentimes, the general contractor will pursue a claim on behalf of the subcontractor. Such an effort may be barred, however, by the *Severin*[66] doctrine. Under this doctrine, a general contractor may only prosecute a claim against the government on behalf of its subcontractor if it has paid the subcontractor's claim or remains liable to the subcontractor for that claim.[67] In other words, the general contractor can only recover from the government if it

---

65. An example of a liquidating agreement can be found in CONSTRUCTION INDUSTRY FORMBOOK 182–85 (CUM. SUPP. Shepard's/McGraw-Hill, 1987).

66. *See* Severin v. United States, 99 Ct. Cl. 435 (1943), *cert. denied*, 322 U.S. 733 (1944).

67. *See* E. R. Mitchell Constr. Co. v. Danzig, 175 F.3d 1369 (Fed. Cir. 1999).

suffered actual damages. Absent such proof of general contractor damages or responsibility to the subcontractor, the government retains its sovereign immunity from suit.[68]

The burden is on the government to prove that the general contractor is not responsible for the costs incurred by the subcontractor that are at issue in the pass-through suit.[69] When the government either fails to raise its *Severin*/sovereign immunity defense at trial, or raises it and fails to prove it, then the claims of the subcontractor are treated as if they are the claims of the general contractor, and any further worry about the absence of subcontractor privity with the government is extinguished.[70]

In order to avoid *Severin* doctrine problems, the liquidating agreement between the general contractor and the subcontractor should not set forth a blanket release of the subcontractor's claim. Alternatively, the general contractor could pay the subcontractor's claim, but that rarely happens.

VII. CONCLUSION

Subcontractors are the reason projects get built since they do most of the work based upon their trade-specific skills. They must, however, protect themselves from beginning to end—in bidding, contract negotiation, contract performance, and collection—in order for their work to be rewarding and profitable. Because subcontractors do not have much leverage against the general contractor, counsel for the subcontractor must be attuned to the legal risks and issues in order to advise the client who will ultimately make the business risk decisions.

---

68. *See id.* at 1370.
69. See id.
70. *See id.* at 1371.

5

A Primer in Contract Surety Law

Charles R. Gossage

I. INTRODUCTION TO SURETYSHIP: WHAT'S THE DIFFERENCE BETWEEN SURETYSHIP AND INSURANCE?

A surety is one who is contractually obligated to answer for the debt or default of another. Sureties, which are almost universally insurance companies, undertake this obligation for a "premium" that is calculated as a percentage of the bonded obligation. Generally speaking, the surety's liability is limited to the amount stated in the bond, which is commonly referred to as the "penal sum of the bond."

Regardless of an insurance company's involvement in the issuance of bonds, a contract of suretyship or bond should not be confused with a contract of insurance or insurance policy. Scores of articles and chapters have been dedicated to explaining the differences between a bond and an insurance policy; therefore, for purposes of this chapter, a simplified explanation will suffice.

A contract of suretyship is customarily a tri-party agreement in which the surety obligates itself, with its principal, to a named obligee. The principal owes the primary obligation to the obligee, with the surety remaining secondarily liable to the obligee. In the event the principal fails to perform or meet its bonded obligations to the obligee, the surety's guarantee may be called upon to carry out the balance of the principal's obligations. Notably, at the time the surety issues a bond, it does not anticipate a loss of any kind. On the contrary, if the surety determines as part of its investigative process, which is commonly referred to as the "underwriting process," that a loss will be incurred, then it will not issue the bond.

An insurer, on the other hand, expects that claims will be made and losses will be incurred in connection with the issuance of policies of insurance. It spreads the risk of such losses over hundreds, if not thousands, of insurance policies by the premium it charges for a particular type of covered loss. Unlike a surety, however, an insurer has the primary obligation to pay for the loss or, in many instances, indemnify its insured for the loss.

II. SURETY'S LIABILITY AND DEFENSES UNDER THE BID BOND

In the world of competitive bidding, which is commonplace on publicly awarded construction projects, but also exists in private construction, sureties are called upon to issue "bid bonds" at the time bid proposals are submitted to the owner. Like most bonds, a surety's rights and liabilities are generally spelled out within the four corners of the bonds, but in some states are dictated by statute.

An owner naturally expects that, if the principal is determined to be the lowest responsible bidder, the principal will enter into the contract with the owner/obligee under the terms of the Invitation to Bid (which includes the terms and conditions of the contract) for the price set forth in the principal's bid. The time, effort, and expense involved in the bid process necessitate that the owner demands some form of assurance from the bidders that the low bidder will execute a contract with the owner. This assurance, or guarantee, comes in the form of a bid bond. The bid bond protects the owner from frivolous bids which may be withdrawn by the bidder if the bidder refuses to enter into a contract (for whatever reason), thereby resulting in damages to the owner from rebidding the work; the cost difference between the low bidder and the second low bidder; and/or any monetary damages resulting from the delay in proceeding with the contract, among other items of expense.

In the event that the principal refuses to execute the contract for whatever reason, the Owner/Obligee can declare the bidder in default and may seek damages from the surety for the total amount of the penal sum.[1] In many jurisdictions, the Owner/Obligee may claim the full penal sum of the bond, notwithstanding that its damages are less than the penal sum of the bond.[2] In other jurisdictions, the Owner/Obligee's recovery is limited to its actual damages resulting from the principal's refusal to enter into the contract, which generally would include the difference between the principal's bid and the next lowest bid, as well as any additional engineering, legal and administrative costs incurred by the Owner/Obligee in rebidding the contract.[3]

In some jurisdictions, by statute, the principal may withdraw its bid provided: (1) the principal can prove that it made a mistake in the preparation of the bid (other than a mistake in judgment as to the scope of the work or the principal's ability to perform the work); (2) prompt notice of the mistake is communicated to the Owner/Obligee; and (3) the principal can prove that there was an error in its bid resulting from a mathematical calculation, such as an error in the extension of unit prices in the contract or a transpositional error in the price of the contract.[4] Notice of the error must

---

1. Customarily, the bid bond is in the amount of 5% of the bid.

2. *See* CAL. PUB. CONT. CODE § 10181, which provides "if successful bidder fails to execute the contract, his bidder's security shall be forfeited to the state"; *see also* City of Merrill v. Wenzel Bros., 277 N.W.2d 799 (Wis. 1979).

3. *Lee*, 73 PA. CONS. STAT. § 1601 *et seq.* defining amount of damages; *but see* Colella v. Allegheny County, 391 Pa. 103, 137 A.2d 265 (1958) (limiting damages to the amount of bid bond).

4. *See, e.g.*, CAL. PUB. CONT. CODE §§ 5100–5109; 73 PA. CONS. STAT. §§ 1601–1608.

be given in strict conformity with the statute, and documentation of the mathematical error must be communicated promptly to the owner, together with such proof as the principal may offer that the error was not one of judgment, but rather one of mathematical calculation. In those jurisdictions that allow withdrawal of the bid under such circumstances, the principal is discharged from liability, and the surety's bid bond is exonerated.

A question that sometimes arises is whether the surety, once having executed a bid bond, is thereafter obligated to issue performance and payment bonds when it has discovered that the bid was improvident or that its principal is incapable of performing the work at the bid price. In such scenarios, the surety must make a decision as to whether it will then "bite the bullet" and take its loss on the bid bond, rather than expose itself to a default by its principal under the performance and/or payment bond, which could result in a substantially greater loss to the surety.[5] In reaching such a decision, the surety must be mindful of the fact that there may also exist liability to the principal, as well as to the obligee under the bid bond. Should the surety refuse to issue the performance and payment bonds, resulting in the forfeiture of its principal's bid, it may, in some jurisdictions, face a claim by the principal against the surety for the principal's lost profit in performing the contract.[6] Hence, the surety should not, without sound justification, refuse to issue a performance and payment bond once it has issued the bid bond.

---

5. Performance and payment bonds are normally written at a penal sum equal to 100% of the contract.

6. American Fidelity Fire Ins. Co. v. Kennedy Bros. Const., Inc., 670 S.W.2d 798 (Ark. 1984) (lost profit recoverable for improperly furnished performance bond).

III. SURETY'S LIABILITY AND DEFENSES UNDER THE PERFORMANCE BOND

While the bid bond provides a limited form and amount of performance guaranties to the owner, the performance bond is generally greater in scope and dollar value to ensure that the principal performs all of its contractual obligations owed to the obligee. Generally, the terms and conditions of the bonded contract must be read together with the performance bond to fully understand and appreciate the surety's potential liability and exposure.

In the absence of specific language to the contrary, a surety has no duty to complete the project or cause the project to be completed in the event its principal defaults; however, the surety may be responsible for all costs to complete the project up to the penal sum of the performance bond.[7] Many bond forms, however, dictate the surety's contractual obligations to the obligee in the event of the principal's performance default. These obligations include, but are not limited to, completion of the project, arranging for completion of the project, financing completion, or payment of a fixed sum of money to the obligee in exchange for a discharge under the performance bond.

The surety has the right to assert its own defenses and all defenses which its principal may have against the obligee's claims, as well as to demand strict observance of the terms and conditions of the bond by the obligee.[8] Hence, the surety has the right to defend on the basis that its principal has performed and discharged its liability to the obligee. It may also defend on the basis that the obligee has not met the terms and conditions of the bond, such as giving timely and proper notice of claim as required under the bond or rele-

---

7. *See* United States v. Henke Constr. Co., 35 F. Supp. 388 (E.D. Mo. 1940); Bill Curphy Co. v. Elliott, 207 F.2d 103 (5th Cir. 1958); Board of Supervisors of Stafford County v. Safeco Insurance Co., 310 S.E.2d 445 (Va. 1983).

8. *Egyptian American Bank, S.A.E. v. United States*, 13 Cl. Ct. 337 (1987), *aff'd* 861 F.2d 728; Spear v. Industrial Comm., 562 P.2d 1099 (Ariz. 1977).

vant statute, or bringing timely suit under the bond as may be required under any limitation of action contained in the bond, or relevant statute.

Upon the issuance of its performance bond, the surety will claim that the bonded contract balance represents the surety's security or collateral for completion of the bonded obligations. Consequently, in the event the obligee (over the surety's objection) pays the principal, the surety may claim a *pro tanto* discharge.[9] Similarly, should the obligee prepay or overpay the principal under the terms of the contract prior to complete performance of the contract, or pay for defective work that must be corrected, the surety may also claim a *pro tanto* discharge.[10]

IV. SURETY'S LIABILITY AND DEFENSES UNDER THE PAYMENT BOND

In the usual contract surety situation, the surety issues a payment bond, sometimes called a labor and material bond, contemporaneously with the execution of the performance bond. Payment bonds are generally required on public work projects to protect suppliers of labor, materials, and other services because of the absence of mechanic's lien rights on publicly owned projects. Similarly, private owners may require payment bonds as security for the payment of work performed in an effort to keep the private project free and clear of mechanic's liens.

Although the payment bond names the same obligee as named in the performance bond (the owner or possibly a general contractor in the case of a subcontractor's bond, as the obligee), the surety's liability under a payment bond runs to and for the benefit of persons supplying labor and material to the prin-

---

9. Central Blacktop Co., Inc. v. Town of Cicero, 519 N.E.2d 972 (Il. App. Ct. 1988), *appeal denied*, 526 N.E.2d 828 (Ill. 1968).

10. Airtrol Eng'g Co., Inc. v. U.S. Fidelity & Guar. Co., 345 So. 2d 1271 (La. App. 1977).

cipal under the bond. Indeed, the named obligee under a payment bond generally has no direct rights against the surety.

The penal sum of the performance bond and the penal sum of the payment bond are, generally speaking, in the same amount. Hence, usually the penal sum of the payment bond should be sufficient to cover all unpaid suppliers of labor and material. Notably, the surety's liability is limited to the penal sum of the payment bond regardless of the total claims which unpaid suppliers of labor and material may have against the principal under its bonded contract. By way of example, if a principal fails to pay a million dollars in unpaid bills to suppliers of labor and material, but the payment bond is issued in the amount of $500,000, the surety's total liability will be limited to the penal amount of $500,000.[11] If the total claims exceed the penal sum of the bond, a cautious surety may file an action for exoneration or interpleader, giving notice to all unpaid suppliers of labor and material to present their claims which, if valid and covered by the payment bond, will be paid pro-rata to the extent of the penal sum of the bond.[12] Such a large number of claims may also give rise to an investigation to determine if the obligee bears some responsibility for such a loss.

As in the case of a performance bond, the surety's liability to unpaid suppliers of labor and material does not exceed its principal's liability, and all defenses which the principal may have to claims under the payment bond are available to the surety. These would include defenses that the principal has paid the claimant in full, or that the performance by the claimant has been defective, or that the claimant's performance has been untimely, resulting in delay damages to the principal. The surety has defenses in addition to those of its principal. As in the case of performance bond liability, the surety may demand strict compliance with the terms and conditions of its

11. See Bill Curphy Co. v. Elliott, 207 F.2d 103 (5th Cir. 1985); Board of Supervisors of Stafford County v. Safeco Ins. Co., 310 S.E.2d 445 (Va. 1983).

12. Hartford Accident & Indemnity Co. v. Wilson, 569 P.2d 308 (Ariz. Ct. App. 1977).

bond and assert as a defense a claimant's failure to meet the terms and conditions of the payment bond.[13] This would include the claimant's failure to give timely notice, as required under the bond, or to bring suit in a timely fashion in accordance with the limitation of action provision of the bond.[14]

The surety may also have defenses under statutory law. The federal Miller Act[15] (and little Miller Act laws of many states)[16] provides that claims must be brought against the surety in a specific manner, after having given required and timely notice of claim when applicable and bringing the action within the statutory limitation period. The statutory requirements are similar to those conditions frequently found in the terms of the bonds. Moreover, the relevant statute of limitations of any given federal or state statute must be observed.

The terms and conditions of the bond affecting the surety's liability to those suppliers of labor and material having privity with the principal differ from those that do not have privity with the principal. Both the standard conditions of bonds and statutes relating to the bonding of public contracts sometimes provide that those making claims under the bond who have no privity with the obligee must comply with additional terms to perfect their claims, such as giving proper notice to the principal and/or surety within a specified time, and limiting the filing of suit within a specified time period.[17] Generally, the terms of the bond and statutes preclude liability of the surety to "third tier" vendors because they are too far removed from the construction process."[18]

---

13. 16 Easton v. Boston Investment Co., 51 Cal. Ct. App. 246 (1921).

14. *Id.* at 249.

15. 40 U.S.C.A. § 270 *et seq.*

16. *E.g.*, 8 PA. CON. STAT. § 191 *et seq.* (Pennsylvania "Public Works Contractors' Bond Law of 1967").

17. *See* 8 PA. STAT. § 194(b) (requiring written notice to contractor within 90 days from date of performing last of labor or furnishing materials as condition precedent for instituting action on bond).

V. SURETY'S RIGHT OF SUBROGATION

Subrogation generally represents an important source of recovery for the surety in the event of the principal's default. Typically, the right of subrogation is asserted by the surety against the performance bond obligee where it has paid and discharged the principal's performance bond liability and the surety as subrogee to the principal demands that the obligee pay the surety all unpaid contract balances.[19] Similarly, where the surety has paid and discharged payment bond liability, it steps into the shoes of those claimants with the right to assert the claimant's rights against the contract balances.[20] The surety's right of subrogation against the contract balances is generally held superior to even those of secured creditors (e.g., trustee in bankruptcy, mortgagee(s) and assignee(s) of contract balances).[21]

VI. UNDERWRITING CONSIDERATION OF THE SURETY BEFORE BONDING THE PRINCIPAL

Underwriting is the process which sureties generally follow before agreeing to write any bond for a principal, including the initial bid bond. Often overlooked, it is important to have a general understanding of this process not only for the representation of sureties, but also to assist contractors unfamiliar with this process.

In general, the surety's local underwriting people make an initial and tentative decision as to whether bonds should be written for a given principal. Before making such a deci-

---

18. *See* Elmer v. United States Fidelity & Guar. Co., 275 F.2d 89 (5th Cir. 1960); Clifford F. MacEvay Co. v. United States; 322 U.S. 102 (1944) (precluding liability to material supplier of a second tier subcontractor).

19. Prairie State Nat'l Bank v. United States, 164 U.S. 227 (1896).

20. Henningsen v. United States Fidelity and Guar. Co., 208 U.S. 404 (1907).

21. *See* Federal Ins. Co. v. Community State Bank, 905 F.2d 112 (5th Cir. 1990) (surety's rights superior to assignee bank); Pacific Indemnity Co. v. Grand Avenue State Bank, 223 F.2d 513 (5th Cir. 1955) (surety's right of subrogation prevailed over bankruptcy trustee); International Fidelity Ins. Co. v. United States, 745 F. Supp. 578 (E.D. Mo. 1990) (relation back doctrine applied allowing surety's subrogation rights to have priority over IRS seizure).

sion, the local underwriters generally require the principal to provide financial statements, a financial history of the principal's company and evidence of construction experience. Many sureties require that such financial statements be certified by a public accountant, although at times, such requirements are relaxed.

The underwriters review the principal's balance sheet to determine the principal's net worth and the quality of the assets, and make some attempt to verify the accuracy of the liabilities. Once having determined that the principal's company is solvent and contains sufficient net worth to ensure solvency during the performance of the contract bonded by the surety, the surety next looks very closely at the principal's "net quick." The term "net quick" means nothing more than working capital.

In those instances where the principal's financial statements are subject to any question, or where the size of the contract may exceed what the surety views to be within the historical capabilities of the principal, the surety will then demand and review financial statements from persons who will agree to indemnify the surety against any loss. Typically, the indemnitors are the principal shareholders, partners, officers and directors of the principal's company. As in the case of the review of the principal's financials, the surety may demand that such financial statements of the indemnitors be certified, although this is frequently waived because of the expense incurred in supplying certified financial statements.

Obviously, during this review process, the surety should make some cursory review of the contract and the liability undertaken by the principal, giving particular attention to the form of the bonds required under the contract documents. Time and expense usually prevents any in-depth review of the contract, but a prudent surety should at least look at the time constraints of performance of the contract, the application of any liquidated damages for late completion and the potential for the surety's liability for such liquidated damages.

In the underwriting process, the surety may also be concerned as to the succession of interest in the principal's business—that is, to whom does the equity interest in the principal's business pass in the event of the deaths of the equity holders and/or existing management? The surety may well ask, is there a buy/sell provision in the principal's partnership agreement (or corporate documents, as the case may be), and if so, to whom will the management of the business succeed? The surety may also want to determine whether successive management will be capable of performing the contract.

Each surety will likely have an underwriting manual which prescribes the review process in underwriting. Some manuals contain very stringent conditions, whereas others may be quite loose. The underwriting manual will also provide the authority of the local underwriter to issue the bond. Quite often, bonds may be issued by the local underwriters in relatively small penal sums only. The amount of authority will vary from company to company.

Where the amount of the requested bond exceeds the authority of the local underwriter, the local underwriter must forward an underwriting report, together with copies of the proposed bond, and the principal's and indemnitors' financials, to either the zone or the home office of the surety for final review and approval. Only after such approval is granted may the bond be written.

There have been cases where bonds have been written without appropriate authority, either by an agent of the company or by the local underwriter who has exceeded his authority, and where the surety has attempted to void the bond by reason of lack of authority of the person or persons who have issued the bond to write the bond. The surety's efforts in this regard are generally unsuccessful in voiding the surety's obligation, unless it can prove that the obligee of the bond knew or had reason to believe that authority did not exist (e.g., based upon the existence of defects on the face of the bond(s)).

Customarily, Powers of Attorney are attached to the bonds granting certain named individuals authority to issue the bonds. Where such Powers of Attorney have been attached, and the persons named in the Powers of Attorney have issued the bond, the surety's efforts to void the bond are virtually hopeless, absent fraud. Conversely, where an obligee accepts a bond without a Power of Attorney, the obligee may assume certain risks in respect to the validity of the bond.

VII. SURETY'S RIGHTS TO CONTRACT BALANCES PAID TO PRINCIPAL AND DIVERTED TO OTHER CONTRACTS NOT BONDED BY SURETY

Often in the cases of default, a surety's principal receives contract balances from jobs bonded by a surety and then diverts these funds to other contracts not bonded by the surety, leaving the surety "holding the bag" in the event of a default. The surety, under some circumstances, may trace such funds into the hands of third parties and recover them. This right may exist even where no trust fund statute has been established.[22]

It should be observed that the surety's right to diverted contract balances also may work against the surety in a reverse situation, as the surety may not claim right, title and interest in and to contract balances received by the principal in non-bonded jobs, nor may it generally apply contract balances received from one contract bonded by the surety to indemnify it against losses under another contract bonded by the same surety.[23] This may be so even where the general indemnity agreement (GIA) provides otherwise.

---

22. The trust fund statutes enhance the surety's position in that contract balances paid to a principal become trust funds for the payment of unpaid suppliers of labor and material.

23. Dependable Ins. Co. v. United States, 846 F.2d 65 (Fed. Cir. 1998).

VIII. SURETY'S RIGHT TO INDEMNIFICATION

Under common law, the surety has a right to indemnification and/or exoneration of its liability from its principal.[24] Unlike a contract for insurance, the principal is not insured but, on the contrary, owes a common law duty to the surety to indemnify or reimburse the surety for any losses the surety sustains. Although this is a valuable remedy at law, it frequently has no practical application because, in most cases where a surety has to respond under its bonds, the principal is by then insolvent or bankrupt. Nevertheless, this right can be beneficial to the surety in bankruptcy.

Moreover, contract performance and payment bonds are normally underwritten only after the principal and indemnitors have executed a GIA with the surety under the terms of which the principal and the indemnitors agree to indemnify the surety against any loss. Although these GIAs vary in terms, they frequently bestow upon the surety a certain security interest in the contract balances, the principal's materials and equipment on the job site, and, at times, a security interest in the indemnitors' personal and real property assets. Indeed, many jurisdictions have held that a GIA is in fact a security agreement which can be perfected against the assets of the principal and indemnitors upon the filing of a UCC-1. Of course, perfection of a security interest gives the surety no greater security interest than is bestowed upon it under state law, and it may be junior in lien priority to superior judgments, liens or security interests.

The law regarding the surety's right to indemnification varies from state to state. Depending upon the language of the GIA, some jurisdictions allow the surety to seek indemnification from the indemnitors only where the surety has actually sustained a loss. Other jurisdictions, however, may allow the surety to collateralize itself from losses before they

---

24. Commercial Ins. Co. of Newark, N.J. v. Pacific-Peru Const. Comm., 558 F.2d 948 (9th Cir. 1977).

are incurred, particularly when the GIA requires the principal and/or indemnitors to post collateral in the event a claim is made and/or reserves are set.

Most often, the indemnity suit is filed as an equity action against the indemnitors for specific performance of the indemnitors' obligations contained in the GIA; these include the indemnitors' duty to place the surety in funds to the extent of the surety's contingent liabilities. The purpose of such an equity action serves multiple purposes. First and obviously, it permits the surety to enforce its rights of indemnification for any losses it sustained under its bond and/or to be placed in funds as security for any prospective contingent liability losses. Secondly, it permits the surety to act swiftly to prevent the principal and indemnitors from dissipating their assets while the surety negotiates or litigates claims advanced against it under its bonds. Finally, and of no less importance, such an equity action gets the attention of the principal and the indemnitors, and may encourage them to cooperate with the surety in the defense of claims advanced against the surety.

IX. CONCLUSION

In issuing bid, performance and payment bonds, a surety provides a valuable guaranty indispensable to public works projects, as well as to private works of improvement when the owner or contractor requires the work to be bonded. In the event a claim is made against a bond, it is paramount to the successful prosecution or defense of such a claim that one be familiar not only with the terms and conditions of the bond, but also those of the bonded contract and all applicable statutes. Further, in the event the attorney represents the principal on the bond, he or she should be familiar with all of the principal's obligations to the surety, contractual or otherwise. Following these basic steps, coupled with an understanding of suretyship, will allow the practitioner to begin maneuvering through almost any claim scenario.

Delivery Systems

Robert A. Rubin
Linda M. Thomas-Mobley

6

I. INTRODUCTION

In recent memory, the assembly of a construction project followed a predictable format. The person desiring a facility or "owner" would hire a designer such as an architect or engineer to design the structure. Once a particular design was thoroughly revised, it would eventually be accepted and the designer would reduce this specific intention to a two-dimensional set of drawings. These design drawings would then be married to a set of written technical specifications that explained, in greater detail, the desired outcome of the project. Ideally, upon completion of these construction documents, several general contractors would price the construction effort, and the owner would decide which general contractor would be allowed the opportunity to turn the two dimensional drawings and specifications into a three dimensional reality.

Each party to this scheme had a relatively clear and distinct role to play. The owner usually obtained financing for the project, purchased the real estate and cleared any zoning issues in the way of the project going forward. The general contractor was responsible for the means and methods of construction and contracting with specialty subcontractors to perform the various phases of the construction, and the architect that designed the original facility generally acted on behalf of the owner during the construction, for purposes of maintaining design intent and other construction administration tasks.

This design-bid-build scheme described above is the traditional approach, and has been used for the majority of the facilities built in the United States. The traditional approach is only one type of delivery system also referred to as a project delivery system, or PDS. A project delivery system is a way of organizing the building of a project.

Although *how* to choose from among the various systems, and the effect of the various systems on risk allocation among the parties, are not within the scope of this chapter, these concerns are important to the construction professional. The goal of this chapter is to shed light on the traditional approach and other schemes used for the delivery of new construction projects.

II. THE ANATOMY OF A DELIVERY SYSTEM

The term "delivery system" refers to that bundle of arrangements that turn the owner's vision of a facility into the reality of that facility. If all of the players involved in the assembly of a facility could read minds and if there were no disagreements, claims or litigation, there would be no need for a well-defined project delivery system. Since this is rarely if ever the case, it is important to understand and carefully choose the various components that make up the delivery system for a particular project.

The idea of organizing the project delivery system into a bundle of components or frameworks is highly useful in understanding the interrelationships of the various parts of a project delivery system. The components of the project delivery systems are like pieces of a jigsaw puzzle which, when taken separately, describe only a piece of the whole. The puzzle pieces must be assembled before one can get a true picture of how the system works and how this system reacts to external stimuli.

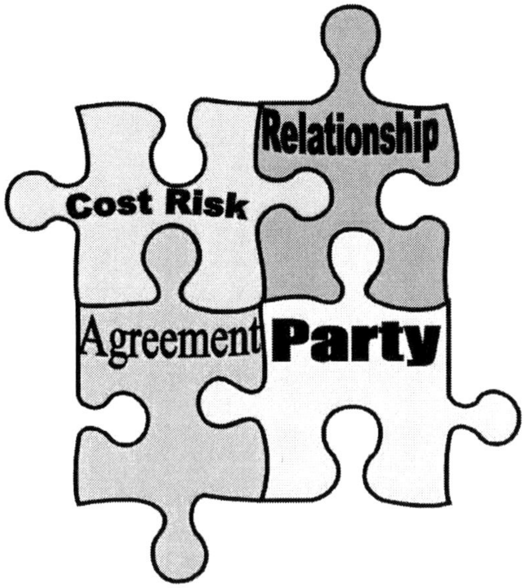

Figure 6.1 Project Delivery System Components

Each delivery system will usually consist of four major components: a financial or cost-risk framework, a relationship framework, an agreement framework, and the Party framework. Depending on which of the various components within a framework an owner chooses, a different delivery system can be obtained.

A. The Cost Risk Framework

The Cost Risk Framework consists of components that describe how the financial risk will be allocated. The most common types of cost risk framework choices are lump sum, cost plus, guaranteed maximum price or unit price.

1. Lump Sum

Lump sum contracts are frequently referred to as stipulated sum or stipulated price or fixed price contracts. The lump sum cost risk framework has been a common agreement in construction projects for the last one hundred years. The contract price is arrived at through a bidding procedure or by negotiations among the parties to the contract. In most cases, though not all, when bids are requested the owner chooses the lowest bid price that meets the technical requirements.

Several assumptions exist under this scheme. In order for a contractor to give an accurate price for the completion of the project, it is assumed that the drawings and specifications are one hundred percent complete and not ambiguous. Often this is not the case.

In a Lump Sum arrangement, the contractor generally bears the risk of cost overruns or other conditions that may require the adjustment of the contract price. In addition, the contractor reaps the benefits of labor efficiencies or materials actually costing less than estimated. This windfall adds to the potential profit of the contractor. Whether a particular adjustment requires an alteration of the contract price is at times hotly debated. Also, in several situations, developers or owners may have changes that are not fully or accurately reflected in the bidding drawings or the written specifications. Seemingly simple changes may have wide reaching and additive effects in the overall cost of a project. Later identified but necessary changes may not be added to the final bidding documents due to time management concerns causing the bidding drawings to be obsolete at the bidding stage. Under this framework, contractors often tend to include contingency costs in the bid price, thereby causing some prices to be artificially elevated.

2. Cost Plus

With the cost plus agreement, the owner is obligated to pay all of the actual costs of the labor, materials and equipment related to the construction, plus a set fee. This fee may be fixed or variable and expressed as a percentage of the total project cost or a lump sum or some combination of the two. Under this agreement, the owner potentially has no control over the final cost of the project and bears the risk of cost overruns and design errors. Also, in a simple cost plus arrangement, the contractor is not required to maintain a cost efficient project since he is assured of a profit regardless of the actual final cost. Moreover, under this agreement, the owner bears the risk of uncertainties. This type of cost risk framework is therefore better situated for small projects that can be contained. Large projects with unknown costs or hard to measure materials, labor or equipment units can expose an owner to great cost escalation.

In order to mitigate potentially large cost escalation, hybrid forms of this arrangement are often used. The contract may provide a cost incentive in the form of a bonus or higher fee if the contractor keeps construction costs under an agreed upon level. These hybrid arrangements are only limited to the contracting parties' imaginations.

3. Guaranteed Maximum Price

The guaranteed maximum price or GMP is somewhat of a hybrid of the lump sum and the cost plus contract. This arrangement is most common in large, sophisticated construction projects. Under this method, the owner and constructor come to an agreement on the maximum price for the project. This guaranteed maximum price includes a percentage profit for the constructor. The arrangement allows for the owner to take less of a risk by setting a maximum for the contract price. As under the cost plus approach, the owner pays the construction costs of labor, materials and equipment and the contractor's profit or fee. In contrast with cost plus approach, the contractor carries the risk of losing all of this profit because he assumes the risk of any cost overruns. In some GMP arrangements, the owner and constructor may agree to share any savings achieved when the project costs are actually lower than anticipated. This arrangement serves as an incentive for the contractor to maintain cost efficient construction practices.

The GMP can be a complicated arrangement to administer. The owner must keep abreast of significant and numerous details. The owner needs the expertise to decipher cost documentation in order to know whether a cost is reimbursable or actually the responsibility of the contractor. Nevertheless, the GMP arrangement may essentially get the owner or project financier closer to the actual price of the project by eliminating the contractor's tendency to add contingency costs to the overall project. In the examples of the contractor saving on labor efficiency or reduced cost of materials, these savings are passed on to the owner if the contractor has agreed to share savings.

4. Unit Price

The unit price arrangement is not commonly used in vertical construction such as commercial or residential building projects. It is more commonly used in horizontal construction such as roads, sidewalks, runways, and infrastructure projects. This method requires the contractor to price his work by unit. A set price will then be paid per cubic foot, square foot or other convenient unit. This is best used when the work is linear or when the full scope of the project is unknown at bid time. Unit prices often include overhead, profit, mobilization and other items not related to any particular items of work. Potential inequitable consequences may result if the anticipated quantities change dramatically. Well-drafted contracts which include a variation-in-quantities clause can be used to balance the risk of drastic quantity decreases or increases.

Cost risk framework decisions are unique to each project. Unfortunately, this framework may be chosen based simply upon which arrangement is most familiar to the owner. It is best to choose the arrangement that is best for the particular project, owner and contractor involved. With the cost risk framework chosen, a relationship framework is then required.

B. The Relationship Framework

The relationship framework consists of components that describe how the major parties involved in the project relate to each other. Relationship framework choices are generally design-bid-build, design-build, construction management (agency or at-risk), and multiple primes.

1. Design-Bid-Build

The design-bid-build approach (the familiar traditional approach) was the principal choice for procurement of construction projects in recent history. Under this relationship framework, when an owner has an idea for a construction project, he first contracts with an architect or designer to devise the scheme of the final facility. The owner may even include in the designer's contract the task of project oversight and project administration. Upon completion of the design, the owner then contracts with a contractor to build the designed project. The relationship that the owner has with the designer is separate and apart from the owner's relationship with the contractor. The contractor as "general contractor" would then enter into separate subcontracts with trade contractors such as HVAC specialists and vendors and material suppliers.

The general contractor has the responsibilities associated with planning, scheduling, and coordination of all the construction activities. Although changing rapidly, this arrangement is still common today.

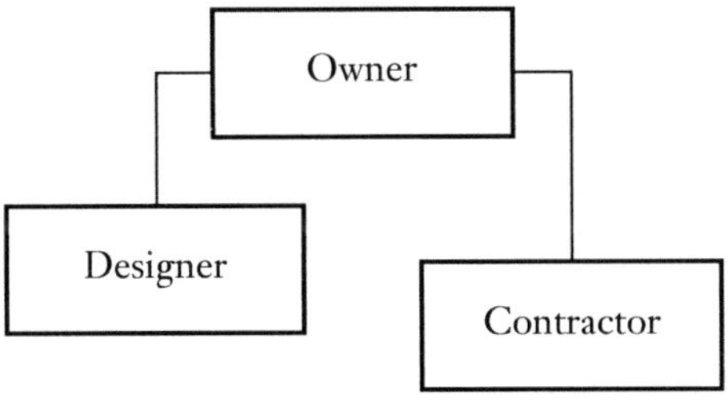

Figure 6.2 Design-Bid-Build Relationship Framework

This arrangement has both advantages and disadvantages. The contractual relationship between the owner and designer, and contractor allow for a sizeable amount of control. The owner has the ultimate approval authority with regard to the final design and the final project as it is constructed. Design intent and quality control may be achieved if the owner elects to have the architect oversee the construction phase. The construction phase services offered by the architect are usually separate from the design services and can prove to be attractive to some owners.

Sometimes, owners will only be involved with a few construction projects over a lifetime and outside expertise is necessary. By its very nature, accountability is difficult in this arrangement. Design contracts can shift the risk for proper construction away from the designer to the owner. Even with construction oversight, the owner makes the final decisions, not the designer. Also, in the traditional arrangement, the architect is not in contractual privity with the contractor, or his subcontractors. This limits the authority the designer may have over the contractor, especially when the contractor is claiming a design error and the designer is claiming a construction error. The owner may end up in the middle of an argument between the designer and the contractor without the necessary technical support to resolve the dispute.

This design-bid-build model may be coupled with the traditional lump sum, the cost plus, GMP, or unit price cost risk frameworks. Each pairing has a different set of concerns and consequences. The choice of which cost risk framework and which relationship framework to use is as distinct as the project and parties involved.

2. Design-Build

The design-build relationship framework has become more popular in recent years, as evidenced by the growth of the

Design-Build Institute of America, an organization committed to the implementation and improvement of the design-build methodology. Under this arrangement, the owner has a single point of contact. This point of contact then arranges for both design and construction services. There are two chief models of this type of relationship framework. In one, the design services are provided for in-house by the contractor; in the other, the contractor has to procure design services. Regardless of which model is utilized, the owner has a single point of contact and a single contract to administer.

The impetus for this approach has been the need for accelerated projects. If the designer and contractor are the same entity, the project time could ostensibly be shortened. Although this may be the case in many design-build schemes, shortened time frames are not always the end result unless specifically agreed upon.

The ease of having one single point of contact is the distinguishing feature of this relationship framework. For the owner, an additional benefit is the cooperation between the designer and contractor. These normally adverse parties are allowed to work together from the beginning, allowing for design changes to occur with greater ease. When design changes are made early enough, the cost impact of the changes is almost always lower. The downside to this arrangement is the loss of objectivity on the part of the designer. The traditional design-bid-build arrangement almost assures a fair amount of tension between the designer and the contractor. When the building entity is also the design entity, contractor profits may drive decisions instead of owner desires.

In addition, the choice of a design-build arrangement may cause an increase in the price of the project because design-build firms are chosen before the design is done. Accordingly, the bidding atmosphere is eliminated, thus reducing competition with regard to price.

There are two additional concerns under this relationship framework. First, the owner is limited to one party from which to recover in the event of a loss. In the traditional method, the owner may have a cause of action against the designer and the contractor, thus spreading the liability.

There are several variants of the design-build approach. One is referred to as the turnkey system. As its name suggests, this procurement system is devised so that once the owner expresses his desire for a new facility and negotiates a contract with a contractor, all he has to do at the end of the project is "turn the key" and his new facility is ready for occupancy. The definition of a turnkey system varies with the parties and may include furniture, fixtures and equipment (FFE) in addition to design and construction services.

Recently, with the move toward outsourcing facility management services, the build-operate-turnover (BOT) project delivery system has emerged. Under this system, the contractor may design, build and operate the facility for sometime prior to turnover to the new owner. This system allows for the facility to be operated over a period of time so that any problems with the technical systems can be identified and resolved prior to independent ownership.

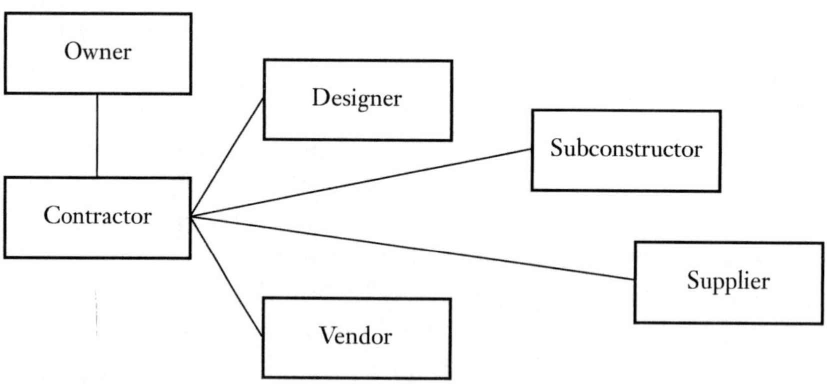

Figure 6.3 Design-Build Framework with External Design

3. Multiple Prime

A less used relationship framework is known as multiple primes. In this method, a sophisticated owner has contractual privity with each trade contractor. This approach is useful for the informed owner who wishes to act as his own general contractor. By contracting directly with each trade contractor and managing all of the coordination and scheduling, the owner may save the general contractor's management fees and can pocket any windfalls gained on the project. One downside to this framework is the lack of a contractual relationship among the various trade subcontractors. This form of relationship also burdens the owner with all of the duties of a traditional general contractor; these duties can be distracting to the owner's everyday responsibilities. Furthermore, in most cases a dedicated on-site coordinator, or construction superintendent, is necessary. Unless the owner has time to remain fully aware of all project details, this arrangement can become too demanding.

The multiple primes relationship framework affords the owner greater control and flexibility. If the project's design changes, the owner has direct dealings with those who will actually execute those changes.

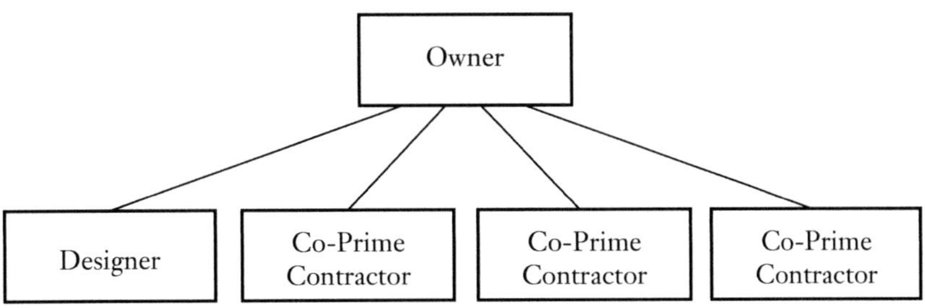

Figure 6.4 Multiple Primes Relationship Framework

4. Construction Management

In the construction manager relationship framework, the owner enters into an agreement with an expert who in turn directs the work. This expert can be a mere advisor (construction manager-agency) who has no contractual relationship with the subcontractors, or an expert who acts as both an advisor and the general contractor (construction manager-at risk). In the construction manager-at risk model, the construction manager has a contractual relationship with the subcontractors and also has full responsibility for construction labor, materials equipment and execution.

a. Construction Management-Agency

The construction manager-agency is often called a "pure" construction manager. His services may also include pre-construction advice. This framework allows the owner to have access to independent advice. In addition, the owner is relieved from maintaining in-house construction expertise.

Construction manager-agency firms review the design documents for errors and constructability issues and the work in place done by the contractor. This manager can mediate disputes between the designer and the contractor in a manner most advantageous to the owner. Finally, the construction manager-agency has no pecuniary interest in the contractor's ability to produce a profit.

On the other hand, the construction manager-agency arrangement can prove more costly to the owner. The fee for this manager is not included in the cost of construction, and regardless of whether the project is successful, a potentially large management fee is due to the construction manager. Also, the general contractor may wish to deal directly with the owner rather than merely through the construction manager-agency firm. Finally, the contract with the construction manager-agency is an additional contract to manage and may create potential liability exposure for the owner.

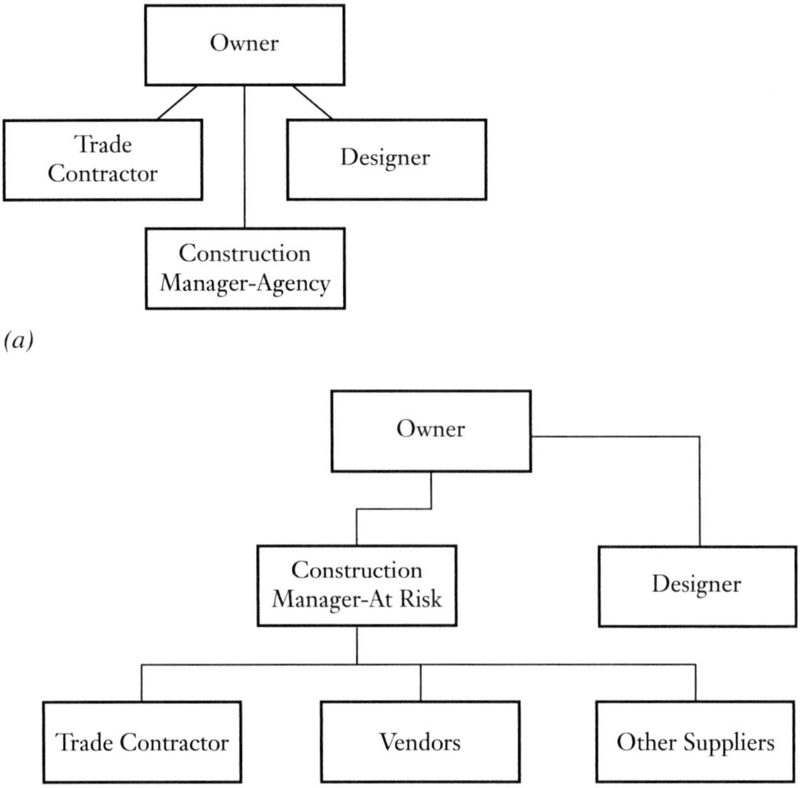

(b)

Figure 6.5 (*a*) Construction Management-Agency Relationship Framework. (*b*) Construction Management-At Risk Relationship Framework

b. At Risk

The construction manager-at risk is another form of this construction manager relationship framework. It should be noted that in the Federal government, the construction manager-at risk relationship framework is referred to as "Construction Manager as Constructor" or CMC. In this scheme, the owner contracts with a construction expert who is *at risk* for completion of the project. The profitability of the manager at risk is connected to the success of the project. Although serving as an incentive for positive results, this potential for loss of profits may cause difficulty for the owner. In some instances, a decision may need to be made which would benefit the owner, but at a significant cost to the contractor. This conflict can test the construction manager-at risk firm's loyalty to the owner. Since it is difficult for the manager to serve two masters, the advice given by the construction manager-at risk cannot be expected to be entirely independent and objective.

The construction manager-at risk model is similar to the traditional design-bid-build approach, where there is a separate designer and contractor, except for one important distinction: the construction manager-at risk is usually involved prior to the construction phase to provide pre-construction services. His advice then continues through to project completion. The continuity of expertise and a full understanding of the owner's concerns expressed at the design stage can often help a contractor understand certain decisions that are made later in the project. Finally, the relationship between the architect, owner and construction manager-at risk can mature earlier during the project. Accordingly, both technical and contractual efficiencies can be realized with this scheme.

C. Agreement Framework

The agreement framework includes components which describe the choices an owner would make concerning the type of written agreement used. The three basic frameworks are found in standard form contracts, such as the forms of the American Institute of Architects, non-copyrighted model forms that suggest terms and conditions, and unique custom agreements written for particular projects. A detailed outline of the construction contract is beyond the scope of this chapter, but it should be noted that the choice of the custom, model or standard form agreement will impact the project delivery system as a whole. Although a project delivery system is not merely a contract, the contract or agreement framework is an integral part of the project delivery system.

Construction industry organizations have formulated copyrighted pre-printed standard form contract documents out of the need for written contracts that could be economically executed and the strong desire to standardize certain types of relationships and practices according to the customs and practices of the construction industry. The various standard form documents are generally promulgated by an organization representing some aspect of the construction industry and, not surprisingly, are protective of the interests of the members of that organization.

Three organizations that develop construction contracts are the American Institute of Architects (AIA), the Engineers Joint Contract Documents Committee (EJCDC), and the Associated General Contractors of America (AGC). The AIA has developed over 75 contract and administrative forms that are in use today. The most widely used standard form in the AIA series is the General Conditions for Construction form. This was the first standard form produced by the AIA in 1911. The 1997 edition (A 201) of this document is the fifteenth edition.

The National Society of Professional Engineers (NSPE) distributes documents developed by the EJCDC. The EJCDC is composed of the American Consulting Engineers Council, the American Society of Civil Engineers, and NSPE's Professional Engineers in Private Practice division. These standardized forms are prepared with the advice of legal counsel and can be used as printed or modified. Modification of these forms should be reviewed by legal counsel prior to use, since many of the coordinating forms are tied into each other.

The AGC develops and revises standard forms via its Contract Documents Committee, composed of contractors, subcontractors, attorneys, insurers and other construction professionals. The AGC's forms are divided into series. The 200 series consists of general contracting documents, the 400 series are design-build documents, the 500 series are construction management documents, the 600 series are subcontracting documents, and the 800 series sort out the Program Management documents. Additionally, the Construction Management

Association of America and the Design-Build Institute of America also produce standard contract documents.

Standard form agreements create a certain level of familiarity among the contracting parties so that contractors and subcontractors know what is expected of them without detailed examination of the construction documents and various agreements. This type of standardization can cause a level of comfort that contributes to a successful project.

One danger, as in any agreement that is not read carefully by the parties prior to execution, is that forms which appear to be the traditional standard form may have been materially altered. The unsuspecting contractor or subcontractor may assume that a project will include a relationship that is familiar to him, bid accordingly, and then be surprised to find that the responsibilities outlined in the modified contract differ significantly from the assumptions made.

D. The Party Framework

The final major element of the project delivery system is the most important element of the system: the parties themselves. The choice of which cost risk framework, relationship framework, and agreement framework the parties will use for a project are all related to the type of parties involved in the project and its execution. A sophisticated owner with many years of experience building a specific type of building may choose completely different components than a first time small facility owner. In addition, the level of expertise and level of involvement of a particular owner may dramatically change a project delivery system with similar cost risk, relationship and agreement frameworks.

III. ASSEMBLING THE PROJECT DELIVERY SYSTEM

The resulting project delivery system is thus related to which components are selected from each of the four different frameworks. In selecting from the cost risk and relationship frameworks, the following grid may be used.

Relationship \ Cost Risk	Lump Sum	Cost Plus	GMP	Unit Price
Design-Bid-Build	A	B	C	D
Design Build	E	F	G	H
Multiple Prime	I	J	K	L
Construction Management-Agency	M	N	O	P
Construction Management-At Risk	Q	R	S	T

Figure 6.6 Cost-Risk Framework and Relationship Framework Grid

The components available in the agreement framework and the party framework are illustrated in the grid below.

Thus, a project delivery system might resemble the following scheme. Design-build and guaranteed maximum price or selection "G" from the first grid, with local government owners using an AIA form contract or selection "22"

Parties \ Agreement	Custom	AIA	AGC	EJCDC	Others
Experienced Owner-Domestic	1	2	3	4	5
Inexperienced Owner-Domestic	6	7	8	9	10
Experienced Owner-International	11	12	13	14	15
Inexperienced Owner-International	16	17	18	19	20
Government-Local	21	22	23	24	25
Government-Federal	26	N/A	N/A	N/A	27

Figure 6.7 Agreement Framework and Party Framework Grid

from the second grid. This project delivery system can be referred to as a "G-22 PDS". For comparison, consider a design-build, lump sum ("E") with an inexperienced domestic owner ("6"). This "E-6" project delivery system is very different from the "G-22" project system. Each has different rules, different sets of responsibilities, and different potential financial exposure for the owner. Yet, in common practice, both of these systems may be called a design-build project delivery system, although they may function very differently.

Not every construction industry organization has every type of form contract document at this time. Moreover, the international construction industry may utilize forms not within the scope of this chapter. The choice of nomenclature may vary from the one chosen by the authors, but the concept remains the same: it is the components of the various frameworks that make up the project delivery system that is of importance, not the name of the PDS. The letter-number shorthand of describing a project delivery system is used here to illustrate to the reader that a construction project delivery system is multifaceted and that merely using the terms, discussed above may not fully reflect all relationships involved in a specific project delivery system.

IV. CONCLUSION

Project delivery systems are just that—systems. The systems consist of four major components that are varied and change depending on the particulars of the project. Since the resulting system changes as the components change, there can be several distinct delivery system combinations. The number of unique systems will increase as the construction industry evolves and as new and innovative cost-risk, relationship and agreement frameworks are developed. It is best when owners are afforded the opportunity to choose from the various components available to them. This choice is based on many different possibilities such as level of sophistication, type of project, resources, need for control and regulatory restrictions and available in-house resources. The construction lawyer should understand all of the individual components and how they relate to each other to form a project delivery system as a whole. This knowledge will go a long way to providing knowledgeable and competent advice to the construction client.

REFERENCES

Robert Dorsey, *Case Studies in Building Design & Construction*,37-48 (1996).

Jimmie Hinze, *Construction Contracts*, 14-25 (2001).

Robert J. Moye III, *Project Delivery Systems: Picking the Right Contracts for Your Client*, Construction Law For the General Practitioner (1999).

Robert A. Rubin & Dana Wordes, *Risky Business*, Nov./Dec., Journal of Management in Engineering, 36 (1998).

Rubin, Fairweather & Guy, Construction Claims, Prevention and Resolution, John Wiley & Sons, 3rd Ed. 1999.

Payment Provisions

L. Wearen Hughes

7

Many construction disputes arise involving the payment provisions in the construction agreement. This chapter will provide a basic understanding of how payment provisions in typical construction agreements are structured.

I. CONTRACT PRICE

A construction agreement typically will contain a provision setting out the total dollar amount to be paid to the contractor (usually called the "contract price" or "contract sum"), or the formula for determining that amount. There primarily are three different types of contracts with respect to such payment provisions: a fixed price contract; a cost plus contract; and a guaranteed maximum price, or GMP contract. Each is discussed below.

A fixed price contract provides for a set total amount to be paid the contractor for the work. The primary AIA document for a fixed price contract is AIA Document A101.[1] Article 4 in AIA Document A101 is the provision dealing with the contract sum.

---

1. Unless otherwise specified, all references to AIA documents will be 1997 editions.

A "cost plus" contract is one in which the contractor is paid for his expenses incurred on the project, plus a fee. A well-drafted cost plus contract will set out in detail what expenses are to be reimbursed and what expenses are not to be reimbursed, as well as the amount of or formula for calculating the fee. Some contracts will call for a fixed fee and some will call for a fee based upon a percentage of the cost of the work.

The third type of agreement, a hybrid of the above two, provides for payment to the contractor on a cost plus a fee basis, but with a "guaranteed maximum price." Under this type of agreement, the contractor is paid based upon his expenses and a fee, but only up to a maximum amount set at the outset by the agreement between the owner and the contractor. The primary, current AIA document dealing with this type of arrangement is AIA Document A111. The contract sum (including the fee) and the guaranteed maximum price are dealt with in Article 5, and the expenses to be reimbursed and not reimbursed are set out in Articles 7 and 8 of that document.

II. PROGRESS PAYMENTS

Good contract documents include explicit provisions governing when and how the contractor will be paid the contract price. Typically, they provide that at stated intervals, usually monthly, a portion of the contract price is to be paid to the contractor, prorated in accordance with the percentage of the total work that has been done.[2] They usually also provide for the owner to retain a small percentage (typically ten percent) of each progress payment until substantial or final completion of the project, which creates a fund called "retainage."[3] The purpose of the retainage is to provide the owner (and the

2. The progress payment provisions of the AIA documents are contained in the respective standard form of agreement (*e.g.*, AIA Document A101, § 5.1 and AIA Document A201 § 9.6).

3. *See, e.g.*, AIA Document A101 § 5.1.6.1.

surety, if there is one)[4] some security against such things as late filed lien claims or incomplete or defective work.

In commercial construction using the AIA documents, the progress payments are based upon a "schedule of values" prepared by the contractor and approved by the architect at the beginning of the project.[5] In the schedule of values, the contractor attributes percentages of the contract price to each of the elements of the work, so that the contract price is divided, pro rata, among the various elements of work according to the cost of each. For example, a $1 million construction contract might have $25,000 allocated to general conditions, $25,000 to clearing and grubbing, $75,000 to foundation and slab, $30,000 to steel framing, $80,000 to electrical work, $80,000 to mechanical work, and other amounts to other components of the work, the total of all these items equaling the contract price.

Typically, the amounts allocated to each part of the work should approximate or equal the amount in the contractor's bid for that item of work. Nevertheless, contractors sometimes "front end load," which means they skew their assigned values in the schedules of values so that work performed early (such as clearing and grubbing, excavation and foundation work) is valued higher than its cost, while later work (such as painting) is undervalued. Front-end loading results in the contractor getting paid more money early in the contract than would otherwise be the case. By front-end loading, the contractor can collect profit before the end of the job; and, if the job has a problem, the owner has less security in the form of unpaid progress payments and retainage than otherwise. The architect should review and, if necessary, object to the schedule of values to prevent this,[6] but may not always do so.

---

4. Equitable principles require that the owner preserve and administer any security the owner holds for the contractor's obligation for the benefit of the surety as well. National Union Indem. Co. v. G.E. Bass & Co., Inc., 369 F.2d 75 (5th Cir. 1966).

5. *See* AIA Document A201 § 9.2.1.

6. *See id.*

The contract usually establishes a day when the progress payments are to be made, usually a set number of days following the receipt, or in some cases, the architect's approval, of the contractor's pay application.[7] AIA Document A201 provides that, ten days before that date, the contractor submits to the architect an itemized (and notarized, if required) application for payment, with whatever supporting documentation the architect may require, showing the amount of work done during the applicable pay period.[8] The application assigns a percentage value to each item of work completed (including stored materials) and the dollar value of that percentage of the contract price allocated to that item of work.

AIA Document A201 provides that "unless otherwise provided" the contractor will be paid not only for materials incorporated in the work, and the labor and equipment to incorporate them, but also for materials delivered to the site, but not incorporated in the work.[9] This provision can be altered so that the owner does not pay for material not incorporated, or to allow the owner to pay for materials purchased but stored elsewhere than the site. Good construction practice, however, usually requires the contractor to have on hand sufficient material to avoid stoppages for lack of materials; thus, important equipment and materials are usually delivered to the site enough in advance of their anticipated use so that the work is not held up. Contractors frequently get discounts from their vendors for quick payment; the owner may benefit from those discounts, if the contract assures the contractor he will have the cash flow to make the payments timely. It thus may make sense for the owner to pay the contractor in sufficient amounts to assure adequate cash flow to pay bills owed by the contractor for stored material.

Under the AIA contract documents, by the contractor's submission of a pay application for work and materials, he

---

7. *See, e.g.*, AIA Document A101 § 5.1.3.
8. AIA Document A201 § 9.3.1.
9. AIA Document A201 § 9.3.2.

warrants to the owner that title to all work, materials and equipment included in the pay application will pass to the owner, no later than the time of payment, clear of liens and not subject to any security interest or encumbrance.[10]

The architect must, within seven days of receiving the application for payment, issue a certification to the owner, based on the architect's evaluation of the work and the pay application, that the work has progressed to the point indicated in the application, that to the best of the architect's knowledge, the quality of the work is in accordance with the Contract Documents (subject to certain exceptions), and that the contractor is entitled to payment.[11] If the architect cannot make those certifications, AIA Document A20l specifically allows the architect to decline to certify payment.[12] The architect may also nullify his certification of a previous pay application, even after the payment is made, if he or she discovers the certification was erroneous, e.g., work previously certified and paid for is defective, or unpaid suppliers file liens; the architect may then reduce the amount of any current pay application to the extent he or she nullified the earlier application.

Moreover, the architect may withhold certification or nullify all or a part of any certificate for payment previously issued, as necessary to protect the owner from loss, because he has discovered: defective work that was not remedied; third-party claims are filed or probable (unless acceptable security is provided to the owner); the contractor has failed to pay subcontractors, laborers or materialmen; evidence that the work cannot be completed for the unpaid balance of the contract sum (i.e., that the contractor had been overpaid for earlier work); damage to the owner or another contractor; reasonable evidence the work cannot be completed within the contract time; or the persistent failure of the contractor to do the work

---

10. AIA Document A201 § 9.3.3.

11. AIA Document A201 §§ 9.4.1, 9.4.2. It is important to note that this certification is not absolute, and is subject to inspection upon substantial completion. It is also subject to certain disclaimers by the architect.

12. AIA Document A201 § 9.5.

in accordance with the documents.[13] In a nutshell, the architect is permitted to deduct or withhold from his certification of a payment, otherwise currently due, an amount sufficient to correct deficiencies or pay third-party claimants so as to insure the owner retains enough money to complete the work and pay claims if the contractor fails to do so.

It is important to note that some architects attempt to force the contractor to agree with and accept the architect's assessment of the amount owed the contractor when it is less than the contractor's pay application by rejecting the pay application and returning it to the contractor, demanding that it be resubmitted in an amount satisfactory to the architect. AIA Document A201 provides only that the architect (within seven days of receipt of the pay application) may either issue a certificate "for such amount as the Architect determines is properly due" or notify the owner and contractor why he is not issuing a certificate.[14] If the contractor and architect cannot agree on a revised amount, the architect must "promptly issue" a certificate to the owner in the amount he can properly certify.[15]

Owners and architects are also sometimes careless about overpaying the contractor, relying largely on the payment and performance bond surety in case there is a problem. The surety, in some jurisdictions, has the right to expect the owner to administer the contract to protect the surety. Accordingly, if the owner overpays the contractor and the overpayment later prejudices the surety, the surety may be released to the extent it was prejudiced.[16] In some jurisdictions, an owner is excused, and the surety is still liable, if the overpayment is made upon the advice of a consulting architect or engineer in good faith. The surety then has direct recourse, however, against the architect or engineer for the its loss.[17]

---

13. *Id.*
14. AIA Document A201 § 9.4.1.
15. AIA Document A201 § 9.5.1.
16. National Union Indem. Co. v. G.E. Bass & Co., Inc., 369 F.2d 75 (5th Cir. 1966).
17. *See, e.g., id.*

III. PAYMENT UPON COMPLETION

Most construction contracts provide for two stages of completion: "substantial completion" and "final completion." Each of these stages is discussed below.

A. Substantial Completion

Substantial completion is usually defined by the contract. The AIA documents define it as being when construction is sufficiently complete, in accordance with the Contract Documents "so the Owner can occupy or utilize the Work or designated portion thereof for its intended use."[18] Some courts have also defined substantial completion in the same general way.[19] The date of substantial completion under the AIA Documents is determined by the architect in a certificate of substantial completion.[20]

At the point of substantial completion, the contractor prepares a list of items to be completed or corrected before receiving final payment, and submits it to the architect; this list is often referred to as a "punchlist." The architect may approve the punchlist or add to it. If the architect determines there are items, whether on the list or not, that should be completed or corrected in order for the project to be substantially complete, the contractor must do so before substantial completion is certified. If the architect decides that the work is substantially complete, he or she prepares a "Certificate of Substantial

---

18. AIA Document A201 § 9.8.1.

19. *See generally* Becherer v. Merrill Lynch, Pierce, Fenner & Smith, Inc., 799 F. Supp. 755, 766 (E.D. Mich. 1992) ("Substantial completion of the hotel meant that the Unit Owners could utilize the property in accordance with its intended use, *i.e.*, a working business ready to be occupied and used by guests and to generate rent"), *rev'd in part on other grounds*, 43 F.3d 1054 (6th Cir. 1995); Aetna Cas. & Sur. Co. v. Butte-Meade Sanitary Water Dist., 500 F. Supp. 193, 198 (S.D. 1980) ("This Court adopts the following definition of substantial completion. '[T]here is substantial performance . . . where all the essentials necessary to full accomplishment of the purposes for which the thing contracted for has been constructed are performed with such an approximation to complete performance that the owner obtains substantially what is called for by the contract.").

20. AIA Document A201 § 8.1.3.

Completion" which establishes the date of substantial completion, usually the date on which the responsibility for security, maintenance, utilities, insurance and similar items shifts from the contractor to the owner, the warranty periods begin to run and, in some states, lien periods begin to run.[21] At that point, the owner pays the contractor the contract price, including retainage, less any amount necessary to correct or complete work that is incomplete or not in accordance with the contract documents.[22]

B. Final Completion

When the contractor notifies the architect that the work is finally complete and submits a final pay application, the architect will make an inspection and, if the work is complete, issue a final certificate for payment, certifying the entire contract balance is due.[23] Before final payment is due, the contractor usually must provide evidence of certain matters. These often include proof that all bills owed by the contractor have been paid, the consent of the surety to final payment and proof regarding insurance.[24]

If final completion is materially delayed through no fault of the contractor, or by late change orders, the owner, upon such certification by the architect, and without terminating the contract, must pay the contractor the balance due for all work completed and accepted, retaining only enough to pay for the uncompleted work.[25]

IV. PAYMENT APPLICATIONS AS INFORMATION SOURCES

A contractor's application for payment can be an information source in assessing claims. While it does not necessarily

---

21. AIA Document A201 § 9.8.
22. AIA Document A201 § 9.8.5.
23. AIA Document A201 § 9.10.1.
24. AIA Document A201 § 9.10.2.
25. AIA Document A201 § 9.10.3.

bind the owner, an approved pay application can be strong evidence binding on the contractor and the architect as to the truth of what is in the application.[26]

An early step for experts and attorneys in analyzing claims is to compare and contrast pertinent pay applications to the assertions of the claim on key issues. One such issue is the state of completion of various parts of the work at various times. Each application typically contains the certifications of the contractor and architect of how much of each part of the work was done as of the end of the relevant pay period. If, for example, the contractor blames late completion of a project on the owner's or architect's delay in approving mechanical submittals, a principal line of defense to that claim would be to establish that other delays, not attributable to the owner or architect, caused late completion. From the pay applications, one might be able to show that earlier critical work, such as slab or steel work, was behind schedule and delayed completion.

The pay application may also show if the mechanical equipment was delivered earlier than the contractor claims, because typically the contractor will request payment for equipment on site, particularly expensive items. In assessing any construction dispute, the pay applications therefore should be reviewed to see if there are logical connections between them and the issues in dispute.

---

26. The contractor's application is, at least, an implicit assertion by the contractor as to entitlement to payment (*see* AIA Document A201 § 9.3.1); and the architect's signature is his certification, based on "his evaluations," that the work has progressed to the point indicated and "to the best of his knowledge, information, and belief" the quality of the work is in accordance with the Contract Documents. AIA Document A201 § 9.4.2.

8 Scheduling for a Construction Project

Richard F. Smith

I. IMPORTANCE OF SCHEDULING

The project schedule provides the timeline blueprint for all construction projects. Every construction activity involves some degree of schedule planning, since the coordination of manpower, equipment, supplies and materials is always required, regardless of the size of the project. On a larger project, the prime contractor's schedule establishes the overall sequencing of events which, if followed, should insure that the project is completed in a timely manner. This chapter discusses the benefits of proper and effective scheduling, the various scheduling techniques used in the construction industry, and how these techniques may be used in assessing responsibility for delay.

A. Minimize Time-Related Costs

A well-planned and fully completed project schedule is necessary for efficient utilization of resources and timely completion of a project. If the contractor is bidding a project on a fixed-price or guaranteed maximum price basis, then there is a financial incentive to complete the project in the most efficient manner and in the shortest period of time possible. Each day of performance requires the commitment of site resources which could be used on other projects. Field overhead and home office overhead expenses associated with a particular project cannot be allocated to other productive uses until the project is completed. Extending performance on a project restricts a contractor's ability to perform other work and impacts its bottom line.

Subcontractors on a project are also dependent upon the prime contractor's project schedule to minimize time-related costs and coordinate work activities. These subcontractors have the same interests as the prime contractor in minimizing the length of time manpower, equipment, field supervision and home office expenses are dedicated to a project.

The owner who has borrowed money to finance the project has a direct financial interest in seeing the project completed and placed in beneficial use in as short a time as possible. In most instances, the owner cannot apply the project to productive use and begin to repay the financing until the project is completed and producing revenue. The longer it takes for the project to be completed, the more time-related costs the owner incurs.

B. Avoid Inefficiencies

A well-planned schedule is needed to efficiently schedule contract work. The on-site labor should be fully utilized at all times, and equipment should remain at the site only as long as needed. The construction schedule allows the contractor to coordinate the work activities, including the various subcontractors and suppliers, to avoid inefficiencies in their performance. For example, a painting subcontractor cannot perform interior painting until the drywall subcontractor has finished its work. With a proper schedule, the prime contractor can coordinate the work to avoid delay between the time of completion of work by one subcontractor and the beginning of work by the next subcontractor in the sequence.

Efficient scheduling of subcontractors and suppliers provides one of the greatest challenges to a prime contractor, since it frequently lacks complete control over them to ensure timely work. The prime contractor's construction schedule informs all subcontractors at the beginning of the project when they will be expected to begin their work and the duration planned for that work. As the schedule is updated, it allows all the parties to adjust their planning accordingly.

C. Timely Ordering, Delivery and Installation of Materials

In addition to manpower and equipment allocation discussed above, scheduling also involves critical issues in the ordering, delivery and installation of materials. Timely ordering and sequencing of delivery of materials prevents the contractor from incurring costs related to premature payment, storage and double-handling of the materials. Many unusual or specialty items on a project may have long lead times from the manufacturer and need to be ordered well in advance of their actual installation at the project. In either case, a schedule which reflects the starting date for an individual subcontractor gives that subcontractor notice of the time period necessary for processing any submittals, shop drawings or other items which might require advance approval prior to ordering the materials.

D. Timely Completion of Project

Owners usually want their projects completed as soon as possible. Most construction contracts will include a "time is of the essence" clause and a substantial completion date for the project. Furthermore, most contracts assess liquidated damages for failure to meet that date. "Substantial completion" is when, under most circumstances, the architect certifies that the project is suitable for its intended use by the owner. The architect certifies "final completion" generally upon completion of all punch list items, normally as a necessary condition to final payment by the owner.

The contract time period usually begins to run when the owner or architect issues a "Notice to Proceed." If the contractor does not substantially complete or fully complete the project within the allotted number of days from the Notice to Proceed, then it may be liable to the owner for a certain amount of money per day under a liquidated damages provision. Occasionally, a prime contractor may have a bonus provision providing a bonus payment for each day that the project is completed ahead of the scheduled completion date or adjusted completion schedule.

It is extremely important for the prime contractor to develop a schedule that allows it to complete the project within the time allowed in the contract. This scheduling process begins well before the contractor's bid is submitted to the owner. The contractor must seek and incorporate schedule input from significant or critical subcontractors or material suppliers to allow them sufficient time to perform their functions or supply materials in an efficient manner so that the project can be constantly progressing. Although the owner is principally concerned about final completion of the project within the allotted time, most owners will also require the contractor to submit updated schedules periodically to verify that sufficient progress for timely completion occurs.

The original schedule often provides a measure for progress payments. A cost loaded schedule allows calculation of progress payments based on the percentage of the completion of project activities. Unfortunately, using a schedule for monthly progress payments often introduces a monthly debate between the contractor and owner concerning percentage of completion. This may change the focus of the schedule from a management tool to a payment tool. As discussed below, an effective schedule is constantly updated and may change depending upon the actual progress of the work and any time extensions allowed by the owner. Schedules are dynamic in nature and must be properly updated to be an effective management tool.

II. TYPES OF SCHEDULING

A. Bar Chart

On smaller projects, a frequently used, inexpensive method of scheduling is the bar chart or "Gantt" chart (named for Henry L. Gantt who pioneered its use in the early 1900s). This type of schedule indicates the beginning and completion dates for various identified activities which make up the work. The length of the bar will indicate the duration of the scheduled work. Within that scheduled period, the chart can also incorporate potential early start dates and/or early finish dates. The bar chart has the construction activities listed in the vertical margin with a time schedule drawn horizontally across the top or bottom of the chart.

While bar charts may be effectively used on small projects with a linear relationship between the activities, they lack detail and do not clearly identify the inter-dependency of various activities. Although a bar chart can be converted to a time-scaled logic diagram indicating some degree of interdependency, a bar chart inadequately describes whether a delay to a particular activity will affect other activities or the ultimate project completion date.

At trial, bar charts require extensive supporting documents and testimony for effective use. Additionally, bar charts have very little use in scheduling manpower, materials and equipment. As a result, most large construction projects will use the Critical Path Method ("CPM") discussed below.

B. The Critical Path Method Scheduling Technique

1. Explanation of "Critical Path"

In the late 1950's, the Critical Path Method of construction scheduling was developed. This method made it possible to display relationships between activities and highlight the interdependencies of the activities which control the progress of the work. It is composed of two parts: (a) a logic diagram depicting all activities needed to complete the work; and (b) the schedule itself. Effective use of a CPM schedule allows a contractor to effectively plan, schedule, coordinate and control the project work activities.

The preliminary steps to prepare a CPM schedule are to list all activities (the individual work items required to be performed) and determine the duration of each activity (the estimated time required to perform the activity). The duration for each activity is normally expressed in workdays. Then the scheduler determines the relationships between activities, i.e., which activities must start before other activities can begin, which activities can be performed concurrently, and which activities must be completed before other activities can begin. See Figure 8.1. The scheduler must consider all factors to determine the time durations and interrelationships, including resources available (manpower, equipment and material), weather, access, holidays, contract required sequencing and contract milestones. Also, the contract may require costs and resource loading where progress shown on the monthly schedule update determines progress payments.

While the schedule can be calculated by hand, it almost always requires the use of a computer to generate the actual schedule, which is based on the logic diagram developed by the scheduler. The schedule utilizes the same horizontal time axis as the bar chart. Instead of bars, activities are depicted by arrows drawn from and to circles which indicate the beginning and ending of that activity. Each activity is time scaled and assigned the length of time based on the

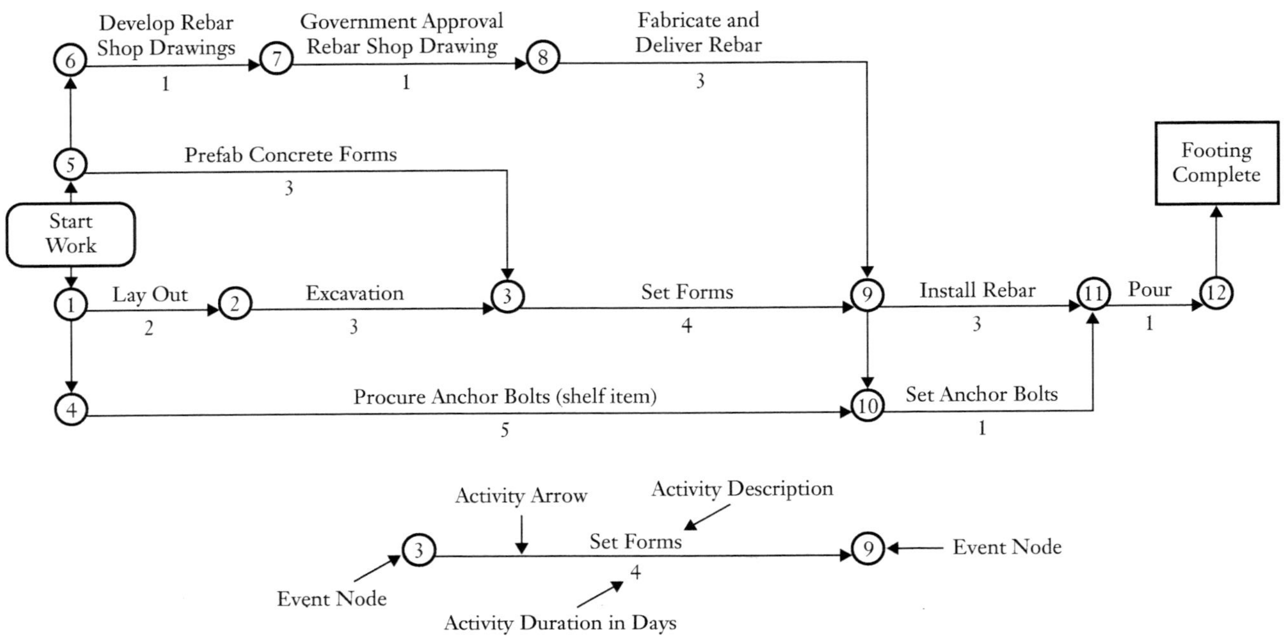

Figure 8.1 CPM Schedule for a Typical Concrete Footing (Typical Arrow Diagram)

work required and the resources available. Using the length of time assigned and the interrelationships established, this type of schedule establishes a "critical path" throughout the project.

This method is called the Arrow Diagramming Method ("ADM") and is also called the I-J or Activity-On-Arrow ("AOA") method, which uses an arrow between two nodes to represent activities. The nodes represent completion of one task and the start of another. Activities are identified by the number of their starting node ("I") and their ending node ("J"). Each I-J number combination is unique to the schedule. All activities with the same "J" node ending number must be completed before activities with the same number as their "I" node can begin.

Using the durations and interrelationships established by the scheduler, the critical path is determined by calculating the earliest time each activity can start ("ES") and the earliest time it can finish ("EF"). This process is called the "forward pass." Then, a "backward pass" is calculated to determine the latest time each activity can start ("LS") and the latest time it can finish ("LF") while still meeting the project completion date.

The critical path is the longest chain of interrelated activities in time through the project from beginning to end. See Figure 8.2. This is the shortest time in which the project can be completed and has no "float" or room for delay. Since this chain of activities will take the longest to complete, each activity on this chain must be completed in the time assigned and is "critical." Any delay in a "critical path" activity without additional effort to accelerate work on the critical path will cause a delay in the final completion of the project. Delays to a subordinate activity (i.e., one that is not on the critical path) may force that activity or some later activities on to the "critical path." As the project schedule is updated, the owner, architect, contractor and subcontractors can identify the activities necessary to keep the project on progress towards a timely completion.

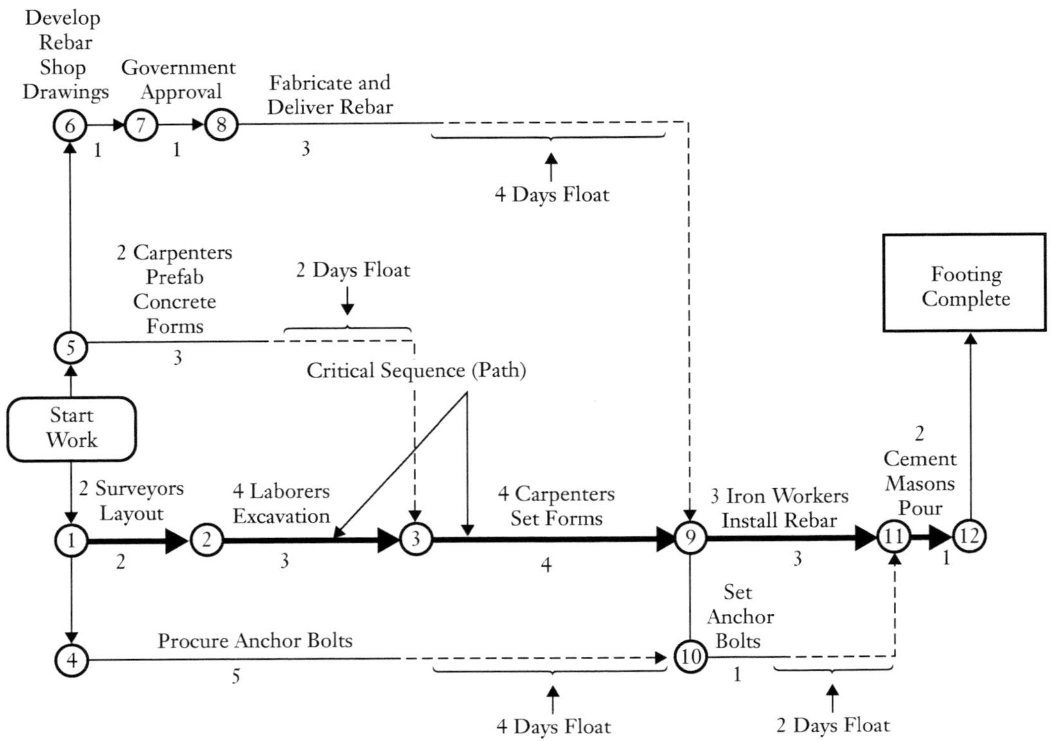

DAY	1	2	3	4	5	6	7	8	9	10	11	12	13
DATE	Mon	Tue	Wed	Thur	Fri	Mon	Tue	Wed	Thur	Fri	Mon	Tue	Wed

Figure 8.2 "As-Planned" Schedule (Time Scaled, Calendar Dated, Critical Sequence (Path) and Float Highlighted)

2. "Float" Concepts and Ownership of Float

On any construction project, more than one activity will be ongoing at the same time. When a schedule reflects that there is more time available to complete a particular activity than that activity requires for timely project completion, the activity is off the critical path, and the additional time is known as "float." Total float is the amount of time an activity can be delayed without delaying the project completion date. Total float is determined along an entire path of activities in the network.

Free float is defined as the amount of time an activity can be delayed without impacting the early start of any succession activity. This is the difference between the earliest and the latest expected times for an event, also described as the difference between the late start and early start, or late finish and early finish. Since that particular activity has the float, a delay in beginning that work will not affect ultimate completion of the project unless the period of delay exceeds the amount of float.

Float allows the contractor the ability to reschedule resources and work without delaying the project. Delays to non-critical activities can push that activity on to the critical path if the delay exceeds the available float.

The ownership of float is often a source of conflict between owners and contractors, because they want to use the float to their advantage. Some owners may assert a contractual right to use the float. The contractor, on the other hand, prefers to control the float to allow flexibility in scheduling. If the owner directs a change which delays a non-critical activity, but sufficient float exists beyond any delay caused by the change to that activity, then the owner rightfully argues that there is no basis for a time extension. However, the contractor may have intended to accelerate the non-critical activity as part of its overall project management, and although the delay does not directly affect the critical path, the contractor will want to exercise its construction management skills in scheduling the work. Alternatively, the contractor may want to delay an otherwise non-critical activity, thereby making it a critical activity. In that case, any change desired by the owner will impact the critical path.

Ownership of float may vary depending upon applicable law and contract terms. While it is ordinarily within the contractor's rights (and obligations) to effectively manage the schedule of the project to insure timely completion, in the absence of a contractual allocation of the ownership of float, the party using the float first generally owns it. For example, if there is ten days of float on a non-critical path and the owner delays the path for eight days and later the contractor delays it for five days, the contractor is responsible for three days of critical path delay. Generally, float can be regarded as an expiring resource which can be used by any party to the contract.

3. How the Critical Path Changes

Changes in the critical path can occur when there is a change in the beginning date of an activity or in the duration of that activity. Any schedule represents the preparer's best estimate of time necessary to complete any particular activity. Differing site conditions, bad weather, labor problems, subcontractor delays, material delivery delays, inefficiencies and other causes, can delay and lengthen the duration for any activity. Additionally, the owner might make changes to the work which either lengthen or shorten the duration of a particular activity.

If the construction activity is on the critical path, then any change in the duration of the activity will change the ultimate project completion date. Most critical path chains of activities include a certain number of days for weather delays built into the activities within the schedule; accordingly, the schedule may frequently have some additional time built in to account for expected weather delays.

As noted earlier, an activity which is not on the critical path because of "float" or "slack" which is built into the schedule may be pushed onto the critical path if there is a delay in starting or performing that activity sufficient to exceed the float. It is the responsibility of the prime contractor to update the schedule as required by the contract to accurately reflect the true critical path and the necessary sequencing of activities.

4. Use of the Schedule

CPM scheduling provides the prime contractor a management tool to plan the work necessary to complete the project. Additionally, the owner or architect will use the construction schedule to monitor the contractor's progress toward timely completion and often to determine progress payments. In the event that a delay occurs, the schedule will become even more important in determining responsibility for the delay and whether the delay will actually extend the project completion date. An updated CPM schedule is the best method available to determine the causes of delay and thus responsibility for time extensions. The "as-built" CPM schedule, recording the actual progress of the work, is an invaluable piece of evidence for a contractor who is preparing a claim for delay damages or for acceleration costs. See Figure 8.3

C. Precedent Diagramming Method ("PDM")

Another networking technique, which is an extension of the Activity-on-Arrow ("AOA") scheduling method, was developed in 1964 by the H. B. Zachery Company in cooperation with IBM. Precedent scheduling has four precedence relationships: finish-to-start, start-to-finish, finish-to-finish, and start-to-start. Precedent scheduling also uses leads and lags. These are the time between the start or finish of an activity and the start and finish of the successor activity.

This PDM method has the advantage of allowing additional precedence relationships of activities rather than

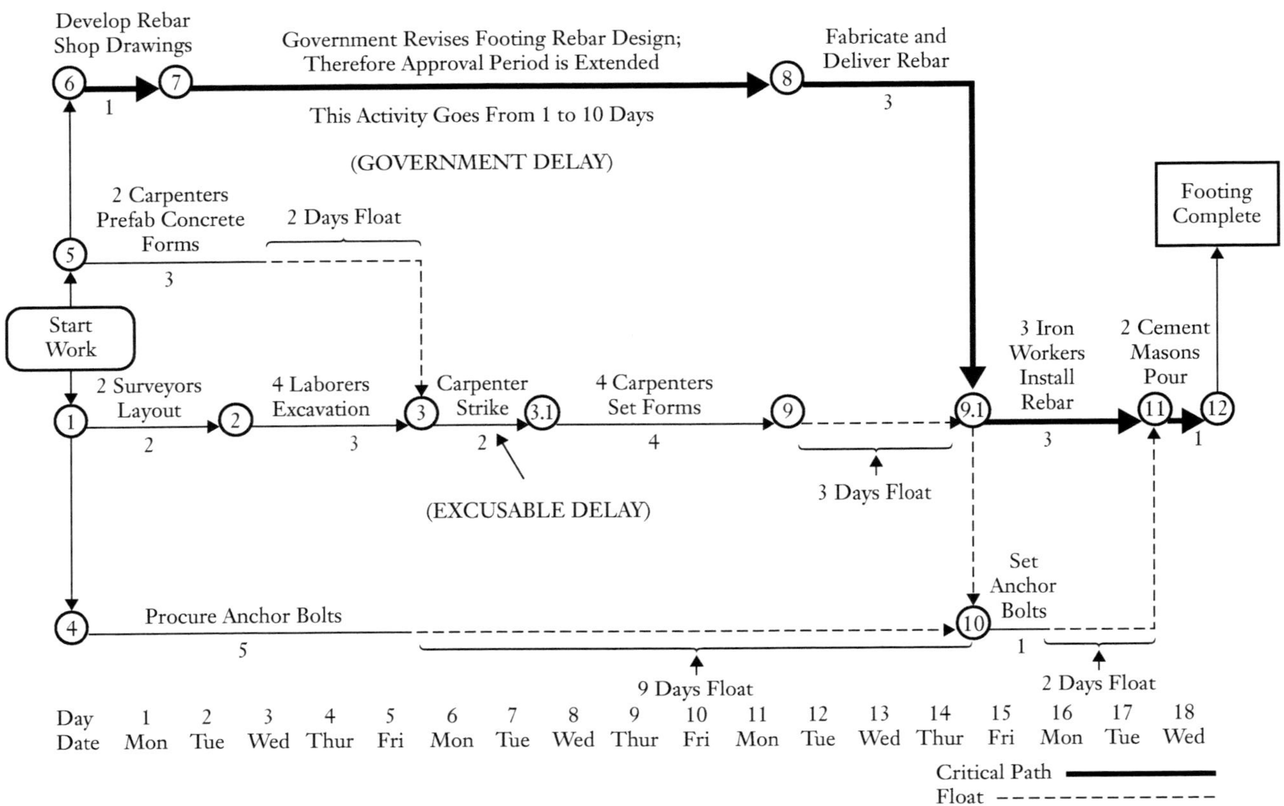

Figure 8.3 "As-Built" Schedule (Critical Path Shifts to Rebar Sequence)

building in dummy activities which are required in the Activity-on-Arrow schedule method. It is more flexible than the AOA method because it defines overlapping and concurrent activities without splitting them into a number of separate activities as required by AOA. PDM is similar to a bar chart by representing activities by boxes and arrows which represent the relationship between activities. It allows an easier development and modification of the network since the logic is independent of the definition of activities.

D. Project Evaluation and Review Technique ("PERT")

In the 1950s, the United States Navy Department and Lockheed, in connection with the development of the Polaris project, developed a scheduling technique know as PERT. This methodology incorporates estimated durations of activities to arrive at an optimistic, pessimistic and most likely range of time for completion. This method is rarely used in construction.

III. DETERMINING RESPONSIBILITY FOR DELAY USING CPM SCHEDULES

Delays which cause a schedule extension can be categorized into three types: those which are excusable without compensation, those which are excusable with compensation and those which are non-excusable without compensation. Excusable delays without compensation are those typically beyond the control and without the fault of either the owner or contractor, e.g., weather delays. Excusable delays with compensation are typically those resulting from the fault or direction of the owner, e.g., a change order. Non-excusable delays without compensation are those resulting from the fault or act of the contractor, e.g., failure to have adequate manpower to perform the work. The contract and case law must be considered to determine how to categorize a delay. Schedule analysis can be used to determine what caused the delay.

A variety of methods using CPM schedules can evaluate delay. All contract provisions concerning preparation of and schedule updates, as well as justification required for delays should be carefully considered in determining the CPM method to use. Existing case law applicable in the jurisdiction should also be reviewed.

Sometimes the contract will specifically indicate how delay should be calculated with respect to method and the time the delay evaluation should be performed. These provisions must be carefully reviewed to choose the appropriate method to prove delay. The choice of delay analysis technique may change the outcome of the delay analysis. The discussion below identifies various methods which have been developed over the years.

A. "Impacted As-Planned" Technique

The "impacted as-planned" technique of CPM analysis uses the original, reasonable, contract "as-planned" schedule as the baseline. Accordingly, it is necessary that the original "as-planned" schedule be reviewed to ensure that it is accurate and logical. If the "as-planned" schedule is produced primarily to negotiate progress payments, it may overstate the amount of work in the early part of the project.

In theory, inserting into the schedule all of the non-excusable delays (contractor delays) and the excusable but non-compressible delays produces a schedule that projects when the work could have been completed "but for" owner-caused delay. Comparing this completion date to the actual completion date provides the claimed amount of compensable (owner-responsible) delay. See Figure 8.4.

Another approach is to insert all of the owner-caused delays into the original, reasonable "as-planned" schedule to demonstrate the owner-caused delay to the contractor's plan to perform the work. A claim then could be made that the resulting extension beyond the "as-planned" completion date constitutes compensable delay. This approach is not

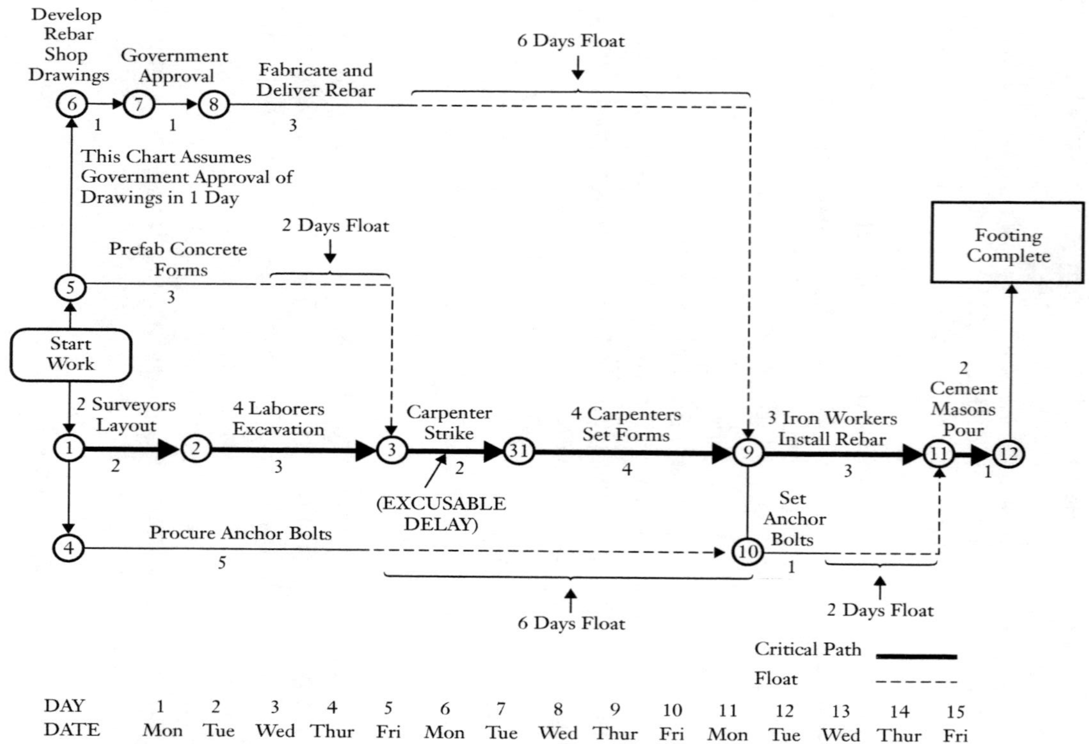

Figure 8.4 "Adjusted" Schedule (Establishing Project Completion Date Absent Government Delays)

generally accepted because it fails to consider the contractor-caused delays and the fluid aspect of a project schedule.

The "adjusted as-planned" method is not favored because of its simplicity. The method generally fails to take into account the sequence and logic changes during the course of the project and the steps taken by the contractor to mitigate the impact of the delay and/or re-deploy its labor and other resources, and contemporaneous logic and resource changes experienced during the work. Accordingly, it has been generally rejected by courts and boards as an acceptable method to prove delay.

B. "Total Time" Technique

The "total time" technique compares the contractor's as-planned schedule with the "as-built" schedule and blames the owner for all delays. This technique is not generally accepted because of the implicit assumptions that the original schedule was complete, reasonable and logical, and that the contractor performed all the work with its originally planned sequence and efficiency. It assumes all delays are the cause of the owner and has been rejected on the same bases as the "total cost" pricing technique.

C. "Fragnet" Schedule Technique

The "fragnet" technique is based on a forward-looking analysis of the delay starting with the status of the project at the time a delay event takes place and a reasonable as-planned schedule thereafter to determine the scheduled completion date at that point in time. The schedule must be properly updated at the time the delay is encountered so the critical path can be determined at that point in time. The "fragnet" approach provides "real-time" analysis of delay events and avoids an after-the-fact analysis at the end of the project.

The first step is to develop a fragmentary network ("fragnet"). The fragmentary network, or fragnet, is a partial schedule apart from the CPM schedule determining the needed new activities, impacted activities, duration changes and logic changes necessary to complete the changed work. The fragnet is generally more detailed than a regular project schedule, but it is limited to just the activities surrounding the changed work.

Once created, the fragnet is inserted into the updated project schedule to create an "impacted" schedule. Comparison of the critical paths on the updated schedule with the "impacted" schedule demonstrates the delay or impact, if any, the change had on the project work. This process is repeated for each delay event encountered during the project.

Many contract CPM specifications, including those of various Federal agencies, require the use of this "fragnet" technology. The "real-time" nature of the analysis does not permit the parties to evaluate subsequent events that may alter the impact of a particular delay, but it does allow the parties to analyze the delay at the time it occurred, considering the plans of the parties at that time. Consequently, the "fragnet" approach may misstate the ultimate impact of particular delays that were either mitigated or extended by later events.

The "fragnet" approach requires that the baseline schedule not contain logic or duration errors and be fully updated at the point the delay event occurs. The baseline schedule must also reflect the way the project is being constructed to accurately reflect the impact of the change at a particular point in time.

Failure of the parties to agree on the "fragnet" and the point in time it should be inserted into the schedule may also cause disputes. If the owner and contractor cannot agree on the work required by the fragnet, they will not agree to the results of the "fragnet" analysis.

The "fragnet" technique can also be used "after-the-fact" to determine project delays, allowing consideration of the overall impact of logic changes or resource use during actual performance of the work. The technique is cumbersome and labor intensive, however, especially where numerous changes or delays must be evaluated.

D. "Windows" Technique

The "windows" technique utilizes "as-built" schedule information as well as "as-planned" schedules. It utilizes a series of schedule updates as chronological "snapshots" to explain major schedule variances over the life of the project. The "windows" technique can also be used after the impact of the delay has been realized, even after project completion.

The "windows" technique focuses on significant project delays. The analysis begins with the selection of one or more periods in time ("windows") to provide a picture of the major delays to the project and their impact to the work. These "windows" are identified through determination of periods in which major schedule variances took place. Once the "windows" are identified, a detailed analysis of that period is undertaken. The "window" technique compares the "as-built" schedule at the beginning of the delay to the updated schedule at the end of the delay to determine the extent of the impact. The technique thus considers what methods or approaches the contractor took to mitigate the impact of a particular delay. Once the first major delay event is analyzed, a second "window" may be created to analyze the next major delay event. This process is repeated for as many "windows" as are considered necessary to analyze the major delays on the project.

Problems arise when the baseline schedule is inaccurate or not properly updated. The parties must also be careful to evaluate the planned schedule for completion of the work to be sure it is reasonable. With the "windows" approach, one must also consider the accuracy of the "as-built" information to insure a proper conclusion.

The "windows" technique focuses on a few major delays, which makes it less cumbersome and time-consuming.

Selection of the particular "windows" may become an issue, however. Since minor delays may be overlooked, there can be issues as to which changes merit analysis and which do not. Using a sufficient number of "windows," this technique can provide an accurate depiction of major delays on the project and provide a balanced prospective and retrospective analysis.

E. "Collapsed As-Built" Technique

The "collapsed as-built" technique utilizes an after-the-fact look at project delays and relies on adequate records to create an accurate representation of how the project was built. The "collapsed as-built" technique creates an "as-built" schedule developed from a detailed review of the project records and interviews with key project personnel. This schedule reflects the actual dates of performance and the actual logic from the records and existing updated schedules.

Each owner-caused delay is then removed or "collapsed" out of the "as-built" schedule with attention paid to the impact on the remaining project activities and the overall schedule logic. Once all of the owner-caused delays are removed, the resulting schedule is reviewed for reasonableness of logic and duration. A comparison of the actual project completion date with the completion date from the "collapsed as-built" schedule results in identification of the delays for which the owner is responsible. The result demonstrates what the schedule would have been "but for" owner-caused delays, and is often referred to as the "but-for" approach.

While the "collapsed as-built" technique is not often specified in contracts, this method has been recognized as an acceptable technique to evaluate delays. It is viewed by some as more reliable due to its foundation on actual performance data. However, determining the "as-built" schedule and quantifying the owner delays to be "collapsed" out requires some subjective analysis similar to that used with the other CPM techniques. Reliance solely upon a retrospective view of the project may ignore the real-time logic change in the schedule to respond to changes or delays as they occur.

The five techniques to analyze delay discussed above are the most common CPM schedule techniques used. Scheduling consultants may present other techniques to analyze changes and delay which are variations of those discussed. These techniques will vary based upon the schedules used as the baseline, the time(s) chosen to focus the analysis (be it prospective, concurrent, or retrospective), the evaluation of concurrent delay, and the method for treating float.

Once the decisions are made regarding activity interrelationships and the allotted time for each activity, use of the CPM theory should provide the same results. The results of a delay analysis can differ, however, depending upon the evaluation technique utilized. Therefore, any CPM schedule analysis must be analyzed with regard to both the accuracy and reasonableness of the underlying decisions and the ability of selected technique to accurately reflect what actually occurred on the project. The contract-specified methodology and case law in the jurisdiction should be consulted as part of the decision process.

F. Concurrent Delay

If the owner is responsible for delay to some aspect of the work (e.g., due to changes in the work) and the delayed activity is on the critical path, the contractor would be entitled to an extension of time, thereby avoiding liquidated damages. Additionally, the contractor may be able to recover damages from the owner for extended overhead expenses and other delay damages. If the delayed activity is not on the critical path, then no basis exists for an extension in the project completion time, although a claim might be made for delay when a unique situation exists, like a piece of expensive equipment was required to remain at the site.

In many instances, the contractor may have some responsibility for causing the delay. When both the owner and the contractor have caused delays in critical path activities, they are considered concurrent delays. Concurrent delay may exist when two or more independent delays occur during the same period of time, any of which would have delayed the project, even if the other delay had not occurred.

Most jurisdictions still follow the rule that neither party can recover from the other where each has contributed to the delays. Under this circumstance, the contractor would be entitled to additional time, but no compensation. Many judicial bodies, however, have struggled with how to deal with current delays, including recovery for compensable delays concurrent with non-compensable delays. Often they will attempt to apportion the delays and expenses attributable to each party, even though the delays are concurrent. The CPM schedule will be a major factor in determining whether the owner's delay or the contractor's delay of a particular activity ultimately caused a delay in completion of the project. If one delay is on the critical path and the other is not, that will almost certainly be the determinative factor.

A unique situation exists when a project is delayed concurrently by the owner and a subcontractor. The general contractor cannot recover from the owner and yet may be responsible to other subcontractors for costs.

IV. ACCELERATION

Acceleration of the work can result if justified time extensions are not granted by the owner, and the contractor is required to expedite the work to meet the unadjusted schedule completion dates. In evaluating a claim for acceleration damages, the courts focus on whether the contractor was ahead or equal to a properly adjusted schedule at the time the order or act requiring acceleration was issued.

Elements of an acceleration claim have developed in the case law and include:

- Excusable or compensable delay as defined by the contract or case law;
- Notice of delay and submission of sufficient support to demonstrate the delay;
- The owner's failure or refusal to grant a time extension;
- Express or implied order by the owner to work on the unadjusted schedule under, for example, a threat to impose liquidated damages; and
- Contractor efforts to speed up the work resulting in extra costs.

Updated CPM schedules help demonstrate what was the proper completion date at the point-in-time of the acceleration order and whether the contractor was on target to meet that date.

V. RIGHT TO FINISH EARLY

Without a contract term prohibiting early finish, most jurisdictions and the federal government support the contractor's right to complete the work earlier than required by the contract. If an early completion is planned, a contractor may claim compensable delay even if it finished before the original contract completion date.

Contractors need to be careful in submitting early completion schedules, since if they are not met, subcontractors may make claims relying on them. Owners have to consider whether to approve an early completion schedule because if they do, the contractor or subcontractors can argue that the owner acknowledged the schedule was reasonable. If the owner rejects the early completion schedule, however, the owner might be deemed to have prevented the contractor from performing the work in its chosen manner. The best tactic for an owner may be to acknowledge receipt of the schedule, without confirming its reasonableness.

Case law recognized elements of a claim for the right to finish early, including conveying the intent to finish early, establishing the capacity to finish early, and proof that the contractor could have finished earlier "but for" the owner's actions. Extended general condition costs may be awarded if a contractor can demonstrate these elements, even if the work was finished before the contract completion date.

VI. THE IMPACT OF SCHEDULE CHANGES

A. Subcontractors

Each subcontractor on a project is dependent upon the prime contractor's schedule to try to schedule its own work. The subcontractor has to anticipate when its work can begin and will need to submit shop drawings, product approvals, samples and other information in time to have those items approved and materials ordered to begin its work at the scheduled time. With more than one project ongoing at the same time, a subcontractor depends upon the general contractor's schedule to try to anticipate its needs for manpower and equipment.

Frequently, the subcontractor will want to use the "float" which exists in the schedule for its benefit in scheduling its resources. Additionally, the subcontractor will need to order sufficient materials to complete the work, but does not want to order those materials in advance of the time when they will be needed (unless it can get paid for stored materials). The subcontractor should also undertake to keep the contractor advised of any changes to the schedule which might be necessary in order to keep the schedule as accurate as possible.

B. Owner

The owner is interested in completion of the project within the time set forth in the contract. Some contracts also have milestone completion dates, and the schedule is used to monitor compliance with those interim dates. The owner has its own schedule for the beneficial use of the project. By following the contractor's updated schedule, the owner can better schedule for itself the probable date of substantial completion and the date on which the owner can put the project to its full use.

Owners need to know when the project will be completed to timely provide the labor to place the facility in use or to rent the facility. Finally, by monitoring the schedule, the owner can make changes in the work which might not affect the ultimate completion date. This involves the use of the "float" in the contract which is discussed above.

C. Architect

The architect frequently has responsibilities which impact the schedule. For example, shop drawings, product submittals, samples and other information must be submitted to the architect for approval. Delays in the approval process can affect the contractor's or subcontractor's ability to meet the schedule. Many contracts include provisions that the architect or owner must act on submittals or shop drawings within a specified number of days, or at least within a "reasonable time." Although the architect's decisions are not usually on the critical path, the architect's failure to act on submittals can force some related activities on to the critical path. If the approved schedule requires drawings to be reviewed within a certain time frame, it may bind the architect to do so.

VII. CONCLUSION

Construction schedules are a frequently changing, constantly updated plan for completion of a project. The general contractor has the responsibility of managing that schedule and the related construction activities necessary to assure timely completion of the project. Properly used, the construction schedule can be an invaluable tool in that process, and can also be an invaluable piece of evidence in any dispute involving the timely completion of the project.

VIII. RESOURCES

A. Articles

Wickwire and Smith, *The Use of Critical Path Method Techniques in Contract Claims*, 7 Pub. Cont. L. J. 1.

Smith and Love, *Scheduling & Proof of Claims, Construction Briefing* 82-6, Federal Publications Inc.

Wickwire, Hurlbut, and Lerman, *Use of Critical Path Method Techniques in Contract Claims: Issues and Developments 1974 to 1988*, 18 Pub. Cont. L.J. 338.

Wickwire and Ockman, *Use of Critical Path Method on Contract Claims–2000*, The Construction Lawyer, October 1999, p. 12.

B. Regulations

U.S. Army Corps of Engineers Regulation ER1-1-11, Progress, Schedules, and Network Analysis Systems, 15 June 1995.

U.S. Army Corps of Engineers Modification Impact Evaluation Guide, EP 415-1-3, July 1979.

Veterans Affairs, VACPM Handbook, H-08-111.

C. Books

The Use of CPM in Construction, Associated General Contractors of America (1976).

Construction Planning & Scheduling, Associated General Contractors of America (1994).

Changes in the Work

Edward Neal Pollard

9

A change is any alteration of a construction contract which changes the scope of the work or adjusts the contract sum or the contract time. Most construction contracts require and permit such changes only by a written "Change Order."[1] The ability to change a contract during performance is essential, for it is the rare project that is completed exactly as initially contemplated.

Changes can be relatively insignificant—as in the color of paints—which likely will not affect contract price or time, or extremely important—as changes in the amount or type of work—which will affect contract price or time.

A good construction contract will have a separate section that deals specifically with how the parties can modify the contract to deal with necessary or desirable changes.[2] When asked for an opinion with respect to changes in the work, or claims arising from changes, a lawyer should first read and understand all the contract provisions that bear on change orders. These provisions will control the

1. AIA Document A201 § 7.1.1.
2. *See* AIA Document A201, art. 7.

debate, if not the outcome, of any negotiation or dispute involving the existence of, or compensation for a change.

I. CHANGES REQUESTED BY THE OWNER

Standard construction contracts contain changes provisions which allow the owner to order changes in the work by written change orders.[3] Without a clause permitting the owner to make changes, the contractor may not be obligated to make these changes. In theory, a change in the work may be viewed as a separate, new job, where negotiations begin at square one. If negotiations break down between the owner and the contractor as to the compensation or extra time required to perform a change, the contractor who can refuse to perform has negotiating leverage, unless the owner has other means (usually another contractor) to perform the work. The owner-directed change clause ensures that for most situations, the changed work progresses. It also protects the project, keeping control of the work from being fragmented. This clause is enforceable, as it was part of the initial consideration on the contract formation.[4] And, most form contracts include methods for calculating a change in price for changes the owner may impose on the contractor, in case they cannot agree on a price.[5]

As contemplated by the AIA standard documents, the process for a scope change begins with the owner or architect defining the change by proposed amendment to the drawings or specifications, and requesting that the contractor offer a proposed price for the change. If the parties agree on the cost and time impact of the change, they simply execute a Change Order, which becomes a contract amend-

---

3. See AIA Document A201 § 7.3.6 and 7.3.7, providing methods for calculating additions to and deletions from the work.

4. Markway v. Kirchenbauer, 769 S.W. 836 (Mo. App. 1989).

5. *See generally* AIA Document A201 § 7.3.6.

ment.[6] The advantage of this procedure is that the owner knows, before incurring additional expense, exactly how the change will impact the cost and schedule of the work.

Most standard form documents permit the owner to order the work, and the contractor must perform, with the price to be computed by an agreed method, usually a cost plus an allowance for overhead and profit.[7]

For contracts with unit prices covering the particular work to be changed, contracts usually allow the owner to order an increase or decrease in quantities at the unit price, unless the magnitude of the change *per se* creates an inequity (*e.g.*, the quantities are decreased to the point that the contractor cannot recoup all his overhead).[8]

Standard contracts also typically require written authorization before the contractor can execute a change in the work, particularly if it impacts contract price or schedule.[9] Recognizing, however, that some jobs operate informally, courts have allowed parties to make oral amendments, notwithstanding contract provisions requiring that changes be in writing.[10]

Some authorities make a distinction between a "change" and an "extra,"[11] although for purposes of the discussion in this chapter, they will be treated as the same. In cases involving contracts that required the contractor to have a written change order before executing a change, some courts have

---

6. *See, e.g,* AIA Document A201 § 7.2.1. With AIA documents, this is done with Document G701.

7. See AIA Document A201 § 7.3.6. Section 7.3.7 provides a method of calculating a price reduction for deletion of work.

8. *See, e.g,* AIA Document A201 § 7.3.3.

9. *See, e.g,* AIA Document A201 §§ 1.1.1 and 4.3.5.

10. *See* Son-Shine Grading, Inc. v. ADC Constr. Co., 315 S.E.2d 346 (1984); Pelican Elec. Contractors v. Neumeyer, 419 So. 2d 1, 4–5 (La. App. 4th Cir. 1982); AA Home Improvements Co., Inc. v. Shane, 217 So. 2d 445 (La. App. 4th Cir. 1969).

11. *See* Somerville and Goldberg, *Changes and Extras* in CONSTRUCTION BUSINESS HANDBOOK, 23–5 to 23–6, suggesting a "change" is an alteration to something already in the work as defined by the contract documents, while an "extra" adds something not in the original documents. *E.g.*, an agreement to dig a well five feet deeper than shown on the plans is a change; and an agreement to dig a second well is an extra. AIA Document A201 §§ 7.2 and 7.3, however, seem to make no distinction between the two.

enforced the requirement;[12] other courts, and this seems to be the modern trend, have allowed recovery based on written confirmation of the change after the work is done.[13]

Indeed, many courts have made the owner pay for extra work that it ordered, without an explicit promise to pay, particularly if there is evidence the owner had reason to expect the contractor to charge extra for the change.[14] Courts have also allowed recovery of a "reasonable amount" for extras, even absent a finding of oral amendment to the contract, on equitable grounds, invoking *quantum meruit* recovery.[15] It is better practice, however, for the contractor, before proceeding, to require both written direction that the work be done, and acknowledgement that the contractor will be paid an additional amount for the change.

II. CHANGES NECESSITATED BY CIRCUMSTANCE

Changes in the work are sometimes imposed upon the parties by circumstances that were not known when they entered into the contract, or have changed since contract formation. Such changes may occur from the discovery of a material difference in a condition from that on which the original price was based, or from some occurrence during the work which changes a material condition. An owner is obligated to pay, and a contractor is entitled to be paid for, costs associated with these changes unless the contractor assumed the risk of the changed condition.

A. Concealed Conditions or Differing Site Conditions

The first of these changes from circumstances are so-called "concealed conditions." Most contract documents address

---

12. *See, e.g.*, Johnson v. Norcion Bros., 95 N.E. 833 (Mass. 1911); Hunt v. Owen Bldg. & Inv. Co., 219 S.W. 138 (Mo. App. 1920).

13. Halverson v. United States, 126 F. Supp. 898 (D. Wash. 1954)

14. Hi Valley Constructors, Inc. v. Ueyser, 428 P.2d 354 (Cal. 1967); C.F. Bolster Co. v. J.C. Boespfly, 330 P.2d 831 (Cal. App. 1958).

15. Villars v. Edwards, 412 So. 2d 122 (La. App. 1st Cir. 1982), *writ denied.*

the possibility that once a contractor begins work, he might encounter some "subsurface or otherwise concealed physical conditions which differ materially from those indicated in the Contract Documents, or otherwise unknown physical conditions of an unusual nature, which differ materially from those ordinarily found to exist and generally recognized and inherent in construction activities of the character provided for in the Contract Documents;[16] for example, the contractor may find a previously concealed condition that neither he, nor the owner, nor the architect, expected. The most common of these are differing underground soil conditions; others include, in renovation or expansion contracts, concealed conditions in the structure to be rehabilitated or enlarged.

Typically, a concealed condition is one that changes the contractor's planned method of construction or adds material, and thus increases the cost. Most contracts require the contractor to notify the owner of a concealed condition immediately after it is encountered, and to obtain the written authorization of the owner before proceeding.[17] The obvious purpose of the notice is to allow the owner and architect an opportunity to investigate, confirm and assess the condition undisturbed, and to explore options, such as changing the design or deleting work, to address the condition. Accordingly, many courts have held that the contractor's failure to notify the owner of such concealed condition within the time specified in the contract waives the contractor's ability to seek payment for additional costs from the owner.

The federal government recognizes, and some commercial contracts have adopted, two different types of provisions dealing with concealed conditions. The first governs when the contract documents affirmatively show a specific subsurface condition, but construction reveals a different condition (a so-called "Type I" changed condition); the second type governs when the contract documents give no information or

---

16. AIA Document A201 § 4.3.4.
17. *See* AIA Document A201 § 4.3.4.

are inconclusive concerning the subsurface condition at issue, but what the contractor encounters is materially different from what experienced contractors familiar with conditions in the vicinity would reasonably expect (a "Type II" changed condition).[18] The A201 document addresses both types.[19] Document A201, § 4.3.4 provides that if the contractor encounters such concealed conditions,[20] the contract sum and contract time "shall be equitably adjusted." The nature and extent of the required notice, and the measure of proof required for a claim related to the condition, are significant sources of contract dispute.

B. Other Changes and Claims

Other events can occur during performance which the contractor believes are contract changes, and can lead to a claim. Some are obvious, as where the architect or the owner gives the contractor costly directives with which the contractor must comply (per § 4.3.5 of A201); others may be less obvious, as where the contractor claims delay damages from the architect's failure to process submittals timely. And some are relatively obscure or uncommon, as when a contractor, whose stored materials were destroyed, discovers the owner failed to purchase the property insurance to cover it, as required by A201, §11.3.1.

In any event, the contractor is required to give written notice of the occurrence of the change leading to a claim immediately upon that occurrence. AIA A201 requires the contractor to give notice within 20 days of the time of the occurrence.[21] This requirement includes claims arising from directives, interpretations or omissions of the architect.[22]

---

18. *See* 48 C.F.R. § 52.236.2. *See, e.g.*, Foster Constr. Co. v. United States, 193 Ct. Cl. 587 (1970).

19. AIA Document A201 § 4.3.4.

20. They must be "concealed" or "unknown" under § 12.2.1; otherwise, the Contractor should have discovered them prior to bidding. *See* AIA Document A201 § 1.2.2.

21. AIA Document A201 § 12.3.1.

Thus, when the contractor gets a directive or interpretation that he thinks might cause a cost increase, the contractor should notify the owner or architect, before incurring any cost based on the directive, that he intends to claim an increase in the contract price or time.[23] The reason is obvious: early notice gives the owner and designer maximum flexibility and options to control cost and damage.

Reduced to its ultimate simplicity, a construction contract, such as the A201, should require that the contractor, if he becomes aware of anything done or omitted by architect or owner which the contractor considers to be violative of the contract on which he will make a claim, to give immediate notice. The notice should be required to be given regarding the condition giving rise to the change. The notice should also state that the change will cause an increase in contract amount and/or time.

III. CONCLUSION

The scope of this chapter does not permit exhaustive treatment of the issue. It is important to remember two things, however, in dealing with any change: first, changes are not, *per se*, unusual—indeed they are to be expected; and second, a good contract will have change and claim clauses which will minimize the uncertainty regarding the parties' rights when a change occurs.

---

22. AIA Document A201 § 12.3.2.

23. Owner or architect directives could include an interpretation of the contract which causes the contractor to perform the work in a way more expensive than he originally had planned. *See* AIA Document A201 § 12.3.2.

Default Termination

Gregory L. Cashion

10

Termination for default is one of the most serious actions which can be taken on a construction project. Before advising about a default termination, an attorney must be familiar with the entire contract between the parties, examine particularly any clauses relating to termination, examine the clauses relating to any dispute underlying the default, be familiar with the facts which justify a termination and evaluate how the dispute between the owner, contractor, surety and others who may be involved will be adjudicated.

The discussion in this chapter focuses primarily on a default termination by the owner. The grounds for termination by the contractor are less prevalent and less complicated. Most standard contract documents require the contractor to keep working as instructed during the pendency of any dispute, while giving the contractor the right to make a claim for a price increase.[1] Rarely does anything but nonpayment justify termination by a contractor.

---

1. *See, e.g.*, AIA Document A201-1997 § 4.3.3.

The greatest risk to either the owner or the contractor in terminating a contract is that the termination could be determined by a court or arbitration panel to be wrongful. If the termination is proved to be wrongful, then the party terminating the contract not only fails to collect its additional funds spent to complete the project, but must also pay the wrongfully terminated party its contract payments through the date of termination and potentially the loss of profit on the work not performed.

I. GROUNDS FOR TERMINATION BY AN OWNER

The contractor has many obligations under a typical construction contract. Whether an event of default exists must, of course, be determined with reference to the contract itself. Most construction contracts, however, define at least the following acts or omissions by the contractor as events of default which entitle the owner, at its option, to terminate: substandard, defective or non-conforming work, failure to pay subcontractors or suppliers, failure to pursue the work diligently, violations of laws, ordinances or regulations, or other substantial breaches of contract. Each of these items is discussed below.

A. Substandard, Defective or Nonconforming Work

The contractor is obligated to perform its work in a good and workmanlike manner.[2] Moreover, the contract will typically provide express warranties regarding the quality of the work and the materials used. In the AIA General Conditions, A201, for example, the Contractor warrants the following:

> . . . materials and equipment furnished under the Contract will be of good quality and new unless otherwise required or permitted by the Contract Documents, that the Work will be free from defects not inherent in the quality required or permitted, and

---

2. *See, e.g.*, Tharpe v. G.E. Moore Co., 174 S.E.2d 397, 399 (S.C. 1970); Weck v. A:M Sunrise Constr. Co., 184 N.E.2d 728, 734 (Ill. Ct. App. 1962).

> that the Work will conform to the requirements of the Contract Documents. Work not conforming to these requirements, including substitutions not properly approved and authorized may be considered defective.[3]

The owner itself may have the authority to reject non-conforming or defective work, or the contract may provide different or additional mechanisms by which such work may be rejected. Under the AIA A201, the architect has the authority to reject work that does not conform to the contract documents.[4] The performance of work by the contractor which is rejected by the architect is a ground for termination of the contract. The owner would have the burden of proving that the defective work is of a material nature to the project and would justify termination of the contract.

B. Failure to Pay Subcontractors and Suppliers

The failure of the contractor to pay its subcontractors and suppliers will expose the owner to mechanic's liens. Most contracts expressly obligate the contractor to "stay current" with all of its subcontractors and suppliers. The AIA General Conditions, A201, provide that the contractor:

> . . . warrants that upon submittal of an Application for Payment all Work for which Certificates of Payments have been previously issued and payments received from the Owner shall, to the best of the Contractor's knowledge, information and belief, be free and clear of liens, claims, security interests or encumbrances favor of the Contractor, Subcontractors, material suppliers, or other persons or entities making a claim by reason of having provided labor, materials and equipment relating to the Work.[5]

The owner may discover the contractor's failure to pay its subcontractors and suppliers through the filing of mechanic's liens against the project, or through less formal communica-

---

3. AIA Document A201-1997 § 3.5.1.
4. *Id.* at § 4.2.6.
5. *Id.* at § 9.3.3.

tions from the subcontractors or suppliers (directly or through intermediaries, such as the architect). The contractor's failure in this regard is a matter to be taken seriously by the owner, and should be investigated as promptly as circumstances allow. The failure of the contractor to pay subcontractors and suppliers is generally a ground for termination of the contract.[6]

C. Failure to Pursue Work Diligently

The contractor has an obligation to pursue its work under the contract in a diligent fashion. Under the AIA A201, the contractor is required to prepare and update a construction schedule which provides for the "expeditious and practicable execution of the Work," and must "perform the Work in general accordance with the most recent schedules."[7] The contractor's failure to pursue the completion of the work in reasonable compliance with the schedule will generally constitute an event of default justifying termination by the owner.[8] The owner should be careful to distinguish between isolated or insignificant deviations from the schedule and serious, ongoing problems with the progress of the work. In that spirit, for example, the A201 provides that the owner may terminate the contract if the contractor "persistently or repeatedly refuses or fails to supply enough properly skilled workers or proper materials" for the project.[9]

D. Violation of Laws, Ordinances, and Regulations

Construction contracts commonly allow the owner to terminate if the contractor defaults by violating laws or ordinances that apply to the owner, the contractor and/or the project. As with the contractor's failure to adhere to the schedule

---

6. *See, e.g., id.*, § 14.2.1.2.
7. *Id.*, §§ 3.10.1, 3.10.3.
8. *See, e.g.*, Black Lake Pipe Line Co. v. Union Constr. Co., 538 S.W.2d 80 (Tex. 1976).
9. AIA Document A201-1997 § 13.2.1.1.

as discussed above, the owner will not be justified in treating an isolated incident as a default justifying termination. Rather, as reflected in the A201, the owner can terminate if the contractor "persistently disregards laws, ordinances, or rules, regulations and orders of a public authority having jurisdiction."[10]

There are, of course, many laws which a contractor could violate which might give rise to the owner's right to terminate. Some examples include failure to pay sales, use or payroll taxes applicable to the work under the contract[11]; failure to obtain permits and license required for the work, assuming that the contract calls for the contractor to do so[12]; and failure to observe safety laws, ordinances and regulations.[13]

E. Other Substantial Breaches of Contract

Most construction contracts should and do contain general provisions which permit the owner to terminate if the contractor commits a material breach of the contract. What constitutes a material breach will depend, of course, on the provisions of the contract. Where the contract is silent or ambiguous, the parties' intent will be inferred by the court, as a question of fact, from extrinsic evidence of the circumstances.[14]

II. ROLE OF CERTIFICATION BY THE ARCHITECT

Depending on the type of default by the contractor and the type of contract drafted by the parties, the architect may have a role in establishing the existence of a default which will permit the owner to terminate the contract. The AIA construction contracts provide a high level of involvement

---

10. *Id.*, § 14.2.1.3.
11. *See, e.g., id.*, § 3.6.1.
12. *See, e.g., id.*, § 3.7.1.
13. *See, e.g., id.*, § 10.2.2.
14. *See, e.g.*, Kuswa & Assocs., Inc. v. Thibaut Constr. Co., 463 So. 2d 1264, 1266 (La. 1985).

for the architect in any effort by the owner to terminate for default. The AIA General Conditions, A201, provide that when any of the enumerated grounds for default exist, the owner may terminate "upon certification by the Architect that sufficient cause exists to justify such action."[15]

Whether the contract requires the architect to certify default or not, the owner should seek the opinion of the architect when considering terminating the contractor for default. In determining the reasonableness of the grounds alleged by the owner for default, the fact finder will place significant weight on the opinion of third parties, such as the architect.

III. NOTIFICATION OF DEFAULT

A. Notice to the Contractor

When the decision is made to terminate the contract because of default by the contractor, the owner must review the contract carefully to determine what notice must first be given to the contractor. Contract provisions which require notice to the contractor and an opportunity to cure are common. Under the AIA A201, for example, the owner must give the contractor (and the surety, if any) seven (7) days' written notice before terminating the contract.[16] Depending on the type of default alleged, a contract may provide specific opportunities for the contractor to cure the default. In the case of defective or non-conforming work, for example, the A201 provides that if the contractor does not commence and continue the correction of the work within seven (7) days of the receipt of written notice from the owner, then

> the Owner may after such seven-day period give the Contractor a second written notice to correct such deficiencies within a

15. AIA Document A201-1997 § 14.2.2.
16. *Id.*

> three-day period. If the Contractor within such three-day period after receipt of such second notice fails to commence and continue to correct any deficiencies, the Owner may, without prejudice to the other remedies the Owner may have correct such deficiencies.[17]

Thus, in the contract language quoted above, as in many other construction contracts, the owner may not terminate the contract without notice and opportunity to cure. In many instances, the contract will allow the contractor a reasonable opportunity to cure the alleged breach and protect its interest under the contract once it receives notice of the owner's alleged basis for termination.

B. Notice to the Surety

Where the contractor has provided a performance bond for the project, the owner must not forget to provide timely notice to the surety before terminating the contract because of the contractor's default. As a general matter, there is no obligation to notify the surety before declaring a default and terminating the contract.[18] A provision in the bond itself which requires notice to the surety, however, will ordinarily be enforced in any subsequent legal proceeding. Moreover, the construction contract itself may require notice to the surety.[19]

Once the owner has complied with the notice provisions in the bond and has declared a default by the contractor, the surety will generally have the option to complete the project itself (directly or through agents) or pay the cost of completion (up to the amount of the bond) to the owner. The AIA document A312 gives the surety the following options:

1. to arrange for the contractor to complete the job, but only if the owner consents;

---

17. *Id.* at § 2.4.1.
18. *See, e.g.*, School Dist. v. Universal Sur. Co., 135 N.W.2d 232, 237 (Neb. 1965)
19. *See., e.g.*, AIA Document A201-1997 § 14.2.2.

2. to perform the contract itself, through agents or independent contractors;
3. to obtain a bid for completion of the work by the contractor acceptable to the owner, secured by bonds equivalent to those issued on the original contract, and pay the owner the cost of so completing the work, up to the amount of the bond;
4. to promptly determine the amount of its liability to the owner, and pay the owner that amount; or
5. to deny liability and notify the owner in writing of the reasons for doing so.

C. Documenting the Default

The notice of default to the contractor and surety is an important piece of information. There is a substantial risk that litigation will ensue after termination, and the notice of default will be a critical piece of evidence in that litigation. Therefore, the notice should be as complete and as objective as possible. In addition, there should be substantial documentation to prove the elements of the default as set forth in the notice to the contractor and its surety.

In the case of defective or nonconforming work, the owner should obtain written reports from the architect and any other parties responsible for monitoring the progress and quality of work, as well as any available photographs and videotapes. If the contractor fails to adhere to the construction schedule, copies of the as-planned construction schedule (including updates) and any available progress reports, such as those contained in the contractor's applications for payment, should be retained in the owner's file for future reference. When the contractor has violated applicable laws, ordinances or regulations, the owner should be sure to obtain copies of any reports or citations issued by the relevant enforcement or regulatory bodies.

Regardless of the default alleged, the owner should bear in mind the potential need to demonstrate the default through documentary evidence in any substantive dispute with the contractor. If possible, a list of persons with knowledge of the surrounding facts and circumstances should also be prepared at the time the decision is made to terminate. As time passes and memories fade, such a list may become a valuable resource.

IV. THE OWNER'S REMEDIES UPON TERMINATION FOR DEFAULT

A construction contract should not only describe the circumstances under which the owner may terminate, but should also contain a list of remedies available to the owner in the event of default by the contractor justifying termination. In the absence of such a list, the court may look to the parties' intent as reflected by extrinsic evidence. It is best to define the owner's remedies in the contract in such a way as to provide the owner the flexibility to complete the project and to protect the owner's interest. The remedies which should be among those included are discussed below:

A. Take Possession of the Site, Unused Materials, Equipment, Machinery and Tools

The owner should be able to take possession of the project in its entirety in order to effect completion by the surety or by a substitute or completing contractor. AIA A201 provides that in the event of default by the contractor, the owner may terminate the contract and:

> take possession of the site and of materials, equipment, tools and construction equipment machinery thereon owned by the Contractor.[20]

---

20. AIA Document A201-1997 § 14.2.2.1.

B. Take an Assignment of Subcontracts or Find New Subcontractors

In almost all circumstances, the work under the construction contract will actually be performed by subcontractors. When the owner terminates the contractor, it may wish to retain some, all, or none of the existing subcontractors to complete the project. Accordingly, the contract should include the flexibility for the owner to take an assignment of the desired subcontracts in the event the general or prime contract is terminated. To create that flexibility, AIA A201 provides that:

> Each subcontract agreement for a portion of the work is assigned by the Contractor to the Owner provide that . . . (1) assignment is effective only after termination of the Contract by the Owner for cause . . . and only for those subcontract agreements which the Owner accepts by notifying the Subcontractor and Contractor in writing; and (2) . . . assignment is subject to the prior rights of the surety, if any . . ."[21]

C. Complete the Project

Once the owner has terminated the contract and excluded the contractor from the site, it probably goes without saying that the owner may undertake to complete the project by whatever means it deems reasonable and appropriate. To ensure that there is no argument over whether the owner must await some adjudication of its claims against the contractor before proceeding with the project, the contract should state that immediately upon termination, the owner may proceed to complete the project as it sees fit.[22]

V. PRACTICAL CONSIDERATIONS

Terminating a construction contract for cause will have serious impacts which go well beyond the owner's legal rights and obli-

---

21. *Id. at* § 5.4.1.
22. *Id. at* § 14.2.2.3.

gations under the construction contract. After a default termination, the contract time and price will almost always dramatically increase. Set forth below are some practical considerations which must be weighed before the owner decides to terminate.

A. Exhaust All Other Options

Many contracts provide the owner with options short of termination simply because termination is such an extreme and risky proposition. The owner should explore those options for the sake of the project, as well as to strengthen its position in any post-termination litigation. The various options should be pursued and systematically exhausted in a diligent and professional manner. Simply going through the motions as part of a rush to termination will not help and could in fact hurt the owner's case. Some of the more common contractual and other options for non-performance include withholding payment; refusal to pay for defective work; liquidated damages; require change of personnel; correct defective work and backcharge contractor; issue a stop work order; require involvement of a surety; issue a deductive change order; declare default, but allow contractor to continue to perform with conditions; declare default and supplement contractor's forces; and renegotiate the contract.

B. Delays in Completion of Project

Perhaps the most obvious practical effect of termination is the resulting delay in completion of the project. For many reasons, the owner may simply not have the flexibility to extend the project schedule to absorb that delay, which may be lengthy and may consist of several components, such as notice and cure period, delays associated with surety's involvement, retaining a completing contractor, procurement procedures with respect to a public contract, and a general loss of momentum. Each of the elements will delay the project, and the owner should evaluate such delays carefully before issuing the notice of termination.

C. Increase in Project Price

Terminating the contract for cause and completing the project with a completing contractor will almost always increase the price of the project for the owner. The increased price may ultimately be recovered in damages from the defaulting contractor and/or its surety, but that recovery will likely occur, if at all, long after the owner has completed the project and has paid the cost of completion. The increased project price can result from several factors. The completing contractor almost certainly will be more expensive, proportionately, for the remainder of the project. The potential for deterioration of the completed work and the risk of loss or damage to stored materials will increase. There will be additional architecture and engineering fees, and perhaps attorney's fees and increased financing costs.

D. Litigation Risks

Finally, there is a risk associated with any threatened litigation. Not only might the owner lose its claim for damages against the contractor, but it may be exposed to a judgment for damages if a judge, jury or arbitration panel finds that the owner wrongfully terminated the contract. In addition, even if the owner is successful in having the contractor determined to be the party in default of the contract, the owner may not recover all of its costs and damages to complete the work of the contractor.

VI. GROUNDS FOR TERMINATION BY A CONTRACTOR

As indicated earlier in this chapter, default determination by a contractor is relatively rare, because in most construction contracts the grounds for contractor termination are few. AIA

General Conditions, A201-1997, §14.1.1, recognizes only three grounds: stoppage of the work by governmental in court action, through no fault of the contractor or any of its forces for more than 30 days; stoppage of the work for 30 days because the architect has not issued a certificate of payment; or failure of the owner to furnish reasonable evidence of financial arrangements to fulfill its obligations under the contract.

The A201-1997, §14.1.3 requires that the contractor give an additional seven (7) days written notice to the owner and architect, after which the contractor has the right to terminate the contract. But most courts have recognized non-payments as grounds for stoppage, and ultimately termination.[23]

As noted above, non-payment by the owner does not necessarily mean failure to pay what the contractor believes is due. Non-payment by the owner means failure by the owner to pay what the architect certifies is due, considering deduction from payments for defective work, lien claims and other contractual grounds for reducing progress payments.[24]

Obviously, on construction projects there will be disputes between the contractor and the architect over the contractor's claims for payment, or the architect's reduction of payments. A good set of contract documents will require a method of resolving those disputes without permitting the contractor to terminate. For example, the AIA General Conditions provide that such claims should be resolved by mediation and arbitration, after the architect has made his decision.[25] The General Conditions do not allow the contractor to stop work, however, during the pendency of the arbitration.[26]

The worse case scenario for a contractor is one in which a dispute arises, the architect reduces or stops payment to the contractor, ostensibly under the specific terms of the contract and, for whatever reason, the contractor has

---

23. United States *ex. rel.* E.C. Ernest, Inc. v. Curtis T. Bedwell & Sons, Inc., 506 F. Supp. 1324 (E.D. Pa. 1981).

24. AIA Document A201 § 9.6.1.

25. *Id. at* §§ 4.5 and 4.6.

26. *Id. at* § 4.3.3.

inadequate resources to continue the work without the payment which has been blocked or withheld. In that circumstance, although the contractor has a duty to go forward, he is without the resources to do so. There are rare and limited circumstances in which the courts have excused a contractor's failure to go forward.[27]

From the contractor's perspective, the contractor should review all available defenses in the event of an owner termination, including at least excusable delays; cardinal change; impossibility of performance; bad faith by the owner; and tortious interference by other or third parties. Finally, with so much at stake in a default termination, all parties should be aware of and be prepared for a bankruptcy filing.

VII. CONCLUSION

Default termination is obviously a very risky endeavor. There are no certain answers on what an owner or contractor should do in a default situation. Each case will likely present many of the problems discussed in this chapter, as well as a few new ones. The mixture of problems will be unique to each case.

The best solution for each case can only be determined with a full understanding of the operative legal principles and, more importantly, a knowledge of the critical facts in the case. In light of the principles and facts, the attorneys for the parties involved should weigh all the factors discussed above, remembering that default is, in most cases, an economic process, and economic considerations will control the decisions.

Special thanks to Charles F. Seemann, Jr. of Deutsch, Kerrigan & Stiles, New Orleans, Louisiana for use of his previously prepared paper on default termination for the American Bar Association, Forum on the Construction Industry.

---

27. Northern Corp. v. Chugach Elec. Ass'n, 518 P.2d 76 (Alaska 1974).

Construction Damages

Andrew D. Ness

11

I. INTRODUCTION AND GENERAL PRINCIPLES

Proving damages is where the rubber meets the road in construction disputes—all the entitlement in the world will not result in recovery if there are not provable damages in dollars and cents. Nevertheless, the principles of proving damages are frequently overlooked or at least under-appreciated; this can easily lead to enormous disappointment in the outcome of a dispute for the party prevailing on the merits. A sound understanding of the principles of proving construction damages is thus a necessity for any construction law practitioner.

Construction disputes are most often contract disputes, and accordingly, the law of construction damages is largely a subset of the law of contract damages. Accordingly, we begin with the general principles applicable to determining damages for breach of contract.

A. Basic Principles of Contract Damages

1. Protecting the Expectations of the Parties

The allowable elements of contract damages are determined from common law contract principles, as established by prior reported decisions. The underlying rationale of contract law, however, is to protect the economic expectations of the parties when they entered into the contract. Accordingly, the law's basic goal in assessing contract damages is to return the non-breaching party to the same relative position it would have been in, absent the breach of contract. In the case of an owner-caused delay, for example, putting the contractor in the same relative position generally means allowing the contractor a time extension for the period its performance was delayed, and compensating the contractor for any additional costs incurred as result of the delay. Because the goal of damages is to remedy the effects of the breach, in proving claims and collecting damages, it is always necessary to demonstrate cause and effect—specifically, that the breach of contract caused the damages sought to be recovered.

The basic principle of restoring the non-breaching party to where it would have been without the breach is, however, subject to important limitations in practice. First, only demonstrable, quantified economic harm will be compensated in damages. In other words, if you have not incurred an amount quantifiable in money, you generally cannot collect it as damages. There is no contract law equivalent to pain and suffering, loss of consortium or similar non-economic damages, as are available under tort law.

Second, not all economic consequences of a contract breach are recoverable. Only those that are considered to be direct and reasonably foreseeable effects of the breach can be recovered. Damages considered remote, unforeseeable or speculative are not recoverable at all, and damages considered as indirect or consequential are subject to significant limitations on recovery, as addressed further below.

2. Preference for Actual Costs over Estimates

In proving damages, actual cost information carries significantly greater weight than estimates of damages. Typical sources of actual cost data include quotations, purchase orders, payroll records, invoices and accounting ledgers. Actual cost records can generally be introduced into evidence as "business records" by someone familiar with how the costs were recorded and maintained in the normal course of the contractor's business. Of course, it is often not possible to utilize actual costs to prove every element of damages. Actual costs may simply not be available, in which case estimates must necessarily be used to prove damages. Estimates are generally considered secondary evidence of damages, but can be sufficient to sustain the claim if properly supported and justified.

Estimated costs will be more readily accepted when actual costs are not available, such as where the contractor's accounting system does not segregate costs for discrete claim items, or when the nature of the work does not lend itself to segregation of actual costs. This is typically the case, for example, with regard to loss of efficiency claims. The use of estimated costs will be viewed much more skeptically where it appears that the claimant simply does not want to disclose actual cost data, or because actual cost information was not maintained in a manner consistent with normal accounting or reasonable business practices.

If estimated costs are used, the weight given to the estimate will depend upon the method or approach utilized in developing the estimate, which method must also be clearly and understandably presented to the fact finder. The evidence must substantiate that the method or approach is a valid means of estimating the costs in question, the calculations made are substantiated and accurate, and the overall resulting estimate is reasonable in amount.

3. Burden of Proof and Standard for Recovery

The burden of proving the validity of the claim, as to both entitlement and amount of damages, is always on the party asserting the claim. The standard of proof which the claimant must meet is "preponderance of the evidence," which means that the claimant must present sufficient evidence to demonstrate that the facts are more likely than not to be as the claimant asserts them. The burden of proof must be met as to all necessary elements of the claim, including damages. Failure to carry the burden of proof as to any one element can defeat the entire claim.

Contract damages can generally be recovered only to the extent that the evidence affords a sufficient basis for estimating their amount in money with reasonable certainty. Damages that are considered to be uncertain, contingent, or speculative typically may not be recovered. Precisely what constitutes "reasonable certainty" in a particular case will vary with the facts. As a rule, mathematical precision is not required, because such precision cannot reasonably be obtained in every case. Often, the nature of the breach itself (such as a series of overlapping delays or hindrances to work) is responsible for the inability to be more precise. Courts generally do not require more certainty than the specific circumstances involved will permit.

When in the normal course of business an element of damages should be determinable from actual cost records, a court is likely to require a higher degree of certainty in the proof of those damages. Many times, however, the actual cost data available does not allocate costs to the discrete items being claimed as damages. For example, an equipment rental invoice may indicate the total rental cost, but often will not show the amount paid while the equipment was idled during a delay. Similarly, labor cost data may not isolate overtime premium pay or the additional labor used to implement a change. In such situations, some uncertainty in the amount of recoverable damages is inevitable, but generally will not prevent recovery so long as a reasonably accurate method of prorating or allocating between recoverable and non-recoverable amounts has been utilized.

In most states, if the court is convinced by the evidence that substantial damages were in fact incurred, and the only uncertainty is as to the amount, the lack of certainty regarding the amount will not bar all recovery. Evidence of the proper damage amount still must be provided, but so long as the fact finder need not resort to speculation to award damages, an amount reasonable under the circumstances may be awarded.

4. Direct vs. Consequential Damages

To be recoverable, breach of contract damages must have been reasonably foreseeable to the contracting parties when they entered into their contract. Clearly foreseeable damages are considered "direct" damages. For example, it is generally considered reasonably foreseeable that a delay in completion of a construction contract will cause damages to both the contractor and the owner of the project.

Indirect or consequential damages are damages that flow from the breach of contract, but are not so obvious and foreseeable as to qualify as direct damages. For example, a contractor who claims that the owner's actions forced it into bankruptcy is claiming consequential damages, as is the owner who claims that the contractor's delayed completion of the project caused it to lose out on new business opportunities. For damages considered as consequential, the claiming party must demonstrate that the claimed damages were a reasonably foreseeable or contemplated result of a breach of this contract at the time the contract was executed. This is a substantial burden in most instances, making recovery of consequential damages considerably more difficult than direct damages, though by no means impossible in most circumstances.

Moreover, liability for consequential damages is often limited or eliminated altogether by specific contract provisions. The current version of the widely-used AIA standard form General Conditions of Contract (A201), for example, now contains a mutual waiver of consequential damages. Such provisions tend to be given effect and enforced, so long as they are clear and unambiguous. Such limitations of liability, however, will not be effective to preclude the recovery of consequential damages in the case of fraud, willful misconduct or (in most states) gross negligence.

B. Damages for Breach vs. Price Adjustment under the Contract

There is a conceptual distinction to keep in mind between breaches of contract, for which the remedy is contract damages, and adjustments to the contract price pursuant to the terms of the contract (such as by a change order issued under the Changes clause). The key difference is that a price adjustment under the contract is not measured by the common law principles of contract damages, but rather by the terms of the contract, which can vary in significant respects from the common law. Since the basic goal of compensating the contractor for the effects of the change is the same, however, either way, the distinction only occasionally becomes a point in dispute.

Similarly, in Federal government and many other public construction contracts (and some private contracts), the term "equitable adjustment" is used to refer to the amount the claiming party may recover for a claim under a remedy-granting clause of the contract. Federal regulations specify what are considered "allowable costs" in equitable adjustments, and cases interpreting these regulations elaborate on what costs may be recovered in such an adjustment. Again, the standards for determining the amount of an equitable adjustment generally parallel the principles of contract damages; the basic goal, that of returning the claiming party to the same relative position it would have been in absent the claim, is the same.

C. The Duty to Mitigate Damages

As a general principle of contract law, the party claiming damages always has a duty to mitigate its damages. This means simply that the claimant must exercise reasonable diligence and ordinary care to avoid unnecessary expenses.

Extraordinary efforts to avoid expenditures need not be taken, but to the extent that a reasonable effort is not made to mitigate damages, recovery will be limited to those damages which would have been incurred despite reasonable efforts to mitigate.

What constitutes appropriate mitigation of damages depends upon the situation. Where the owner has abandoned the project, appropriate mitigation may well require laying off employees, returning rented equipment, cancelling material and equipment deliveries and reselling unused materials where practicable. On the other hand, where work has merely been interrupted by a delay of uncertain duration, appropriate mitigation measures will likely be considerably more modest, such as transferring idle resources (personnel and equipment) to other projects, to the extent practicable. Delays of unknown duration generally do not require discharging personnel because of the costs and uncertainties involved in rehiring them. Even though the delay may prove very lengthy, if the length of the anticipated delay was not known at the time, the contractor cannot be faulted for not taking actions that would have been appropriate had the overall length of the delay been known from the start.

II. DAMAGES FOR CONSTRUCTION DELAYS

Delays to a construction project are one of the most common types of construction claims; as a result, the legal principles relating to delay damages are particularly well developed. Delay damages from the standpoint of the owner and the contractor (or subcontractor) vary considerably, so are addressed separately below.

A. Contractor's Actual Delay Damages

Contractor costs potentially recoverable in a delay claim fall into a few categories. These include the cost of idled labor and equipment; extended project (field office) overhead;

labor, material and equipment escalation; and extended or unabsorbed home office overhead.

1. Labor and Equipment Standby Costs

One of the more common effects of a construction delay is that labor and equipment are either under utilized during the delay period, or idled altogether. Where labor or equipment cannot be productively used during a compensable delay period and must be kept on a "standby" status, the claimant generally will be entitled to recover the actual costs expended for such labor and equipment, presuming these costs could not have been reasonably mitigated.

Recoverable costs for idled labor include the salary and all associated fringe benefits for supervisory personnel, and wages and fringes (labor burden) for hourly workers. The amounts of such costs can usually be determined without undue difficulty from time cards, payroll records and other actual cost records. Recovery for idled labor may be limited, however, by the mitigation requirement that employees must be reassigned to other work during the delay period, to the extent that it is economically sensible to do so.

Damages for idled equipment can be considerably more difficult to determine than for labor. Where equipment is rented from an unrelated party, the actual rental costs paid will be recoverable, provided that recovery will generally be limited to no more than the fair rental value of the equipment involved. Such actual rental costs can usually be obtained readily from invoices or the applicable rental agreements.

The more difficult situation is presented by equipment owned by the contractor. Determining the actual hourly, daily or monthly cost of owned equipment can be difficult or impossible, as many contractors simply do not maintain cost records that allow for such a determination. In some cases, the contract may specify that a particular equipment rate schedule or guide (such as the Corps of Engineers or Associated General

Contractors rates) is to be used in price adjustments under the contract, and this measure can be reasonably extended to damages for breach of contract as well.

Expenses of operating a piece of equipment are not incurred while equipment is idled. Therefore, recovery for idled equipment is limited to the costs of ownership only. Recoverable equipment ownership costs include depreciation, replacement cost escalation, interest on investment, taxes, insurance and storage. Where equipment rates are obtained from an equipment rate guide instead of being based on actual costs, only a percentage of the operating rates in the guide (often 50%) can be recovered. Some guides contain specific factors for determining ownership expense, which should be utilized where available.

Some contractors in their normal course of business develop their own internal equipment rates, and, for cost accounting purposes, charge their projects at those rates for the use of owned equipment. Such internal equipment rates can often be successfully utilized in determining damages, but may be challenged if the internal rates generate an overall profit for the contractor or otherwise do not accurately reflect the contractor's overall costs for owned equipment.

Where a published equipment rate guide is utilized, be aware that the rates for a given piece of equipment in different guides can vary, sometimes significantly. In any event, care must be taken to apply the published rate data correctly and to pay attention to the year or years for which the rates apply. In most rate guides, the published numbers are just guidelines, and may have to be adjusted, using the adjustment factors set forth in the guide, to reflect the particular circumstances involved.

2. Extended Project Overhead

When the duration of a construction project is extended, the costs incurred by a contractor or subcontractor to maintain its presence at the jobsite and continue supervision of the work are extended as well. These costs are variously referred to as "extended project overhead," "extended site overhead" or "extended field office overhead." Extended project overhead is a classically recoverable cost element in a delay claim.

The degree to which extended project overhead is recoverable, however, depends on the extent to which time-related or "fixed" costs are involved. Time-related costs are those costs that continue to be incurred each day, week or month that the project is ongoing. Typically, time-related project overhead costs will include items such as the salaries and benefits of salaried supervisory personnel and field office support personnel (such as timekeepers, accountants and secretaries), costs for site security, and the costs of the field office trailers, storage trailers, and associated utilities, office equipment and telephones. One-time charges that are generally included in project overhead or "General Conditions" accounts, such as bond premiums, insurance premiums, and the costs of initial mobilization, are not time-related. Therefore, they normally are not damages recoverable in a delay claim, since they were not increased as a result of the delay.

The recoverable amount of extended project overhead is typically calculated using one of two methods. The first method is simply to determine from the contractor's accounting records the time-related project overhead costs actually incurred during the claimed delay period. The second method is to total the time-related project overhead costs for the entire project duration, and divide that total by the project duration in days. This result is then the daily overhead rate, i.e., the cost of project overhead per day. The daily overhead rate is then multiplied by the number of days of delay to determine the recoverable extended project overhead.

3. Extended or Unabsorbed Home Office Overhead

Home office overhead refers to the costs of the contractor's home office, including the office rent, salaries and benefits paid to corporate officers and office staff, and costs of centralized services such as a central computer system, centralized accounting services and the estimating department. Home office overhead costs, like most project overhead costs, will typically continue to be incurred at the same fixed rate despite the fact that a project's duration has been extended. Unlike project overhead, however, home office overhead is generally considered an *entirely* time-related cost, since it varies very little, or not at all, with short-term fluctuations in the contractor's workload.

Understanding conceptually why home office overhead can be recovered as delay damages is somewhat difficult; in fact, it has tripped up several courts in reported decisions. Home office overhead is generally not *increased* during a delay, and the delay therefore does not normally cause the contractor to actually spend any more money for home office overhead than if the delay had not occurred. Owners frequently dispute a home office overhead claim for exactly this reason. The damages represented by home office overhead are not premised on the theory that additional costs have been incurred, however; they are instead based on the premise that, because of the delay, other projects must bear a disproportionate share of the home office overhead, and the contractor has been deprived of the opportunity to obtain other projects to share in carrying the home office overhead. Put another way, home office overhead is not fully "absorbed" during a delay period because the direct cost base against which the overhead must be distributed is decreased as a direct result of the delay. Accordingly, recoverable home office overhead is more properly referred to as "unabsorbed," rather than "extended," for delay damages purposes.

Several different methods of calculating unabsorbed home office overhead are possible and justified as a matter of accounting theory. One particular method, however, known as the *Eichleay* formula after the name of the case in which it was first applied, is far and away the most commonly utilized for this purpose. Its popularity has little to do with theory and everything to do with practicality. Although the *Eichleay* formula has been criticized on numerous occasions as not being as theoretically sound as other approaches, it has the incalculable advantage of having been accepted by a substantial number of state and Federal courts.

The *Eichleay* formula's approach to calculating extended home office overhead involves allocating an appropriate portion of the contractor's total home office overhead to the delayed contract (calculation "a" below). The home office overhead allocable to the contract is then transformed into a daily home office overhead rate by dividing it by the contract performance period (calculation "b" below). The daily overhead rate is then multiplied by the number of days of delay to determine the allowable extended home office overhead (calculation "c" below).

a. $$\frac{\text{Contract Billings}}{\text{Total Billings for Actual Contract Period}} \times \text{Total Overhead Incurred During Contract Period} = \text{Overhead Allocable to the Contract}$$

b. $$\frac{\text{Overhead Allocable to the Contract}}{\text{Actual Days of Contract Performance}} = \text{Daily Overhead Rate}$$

c. $$\text{Daily Overhead Rate} \times \text{No. of Days of Compensable Delay} = \text{Recoverable Extended Overhead}$$

A number of cases have criticized the *Eichleay* formula, particularly where it was offered without additional evidence to support the reasonableness of using the formula in the particular circumstances presented. The New York courts, for example, have consistently rejected the use of *Eichleay* in the absence of any proof of actual overhead damages. By contrast, for Federal government contracts, use of the *Eichleay* formula is not only accepted, but has been deemed the *only* acceptable method for determining unabsorbed overhead.

The *Eichleay* formula should not be utilized in a vacuum; it is important to also substantiate the assumptions inherent in the *Eichleay* formula. For example, the contractor will probably need to demonstrate that the home office overhead costs incurred during the delay period could not have been absorbed by taking on other work, because of the uncertain duration of the delay, and that key work elements were suspended and not just stretched out by adding more work (thereby generating a bigger cost base to absorb overhead).

4. Labor and Material Cost Escalation

In an inflationary environment where prices increase over time, a construction delay may cause the contractor's costs for labor, material and equipment on the project to be increased, or "escalated," because they are incurred in a time period later than anticipated. In a proper case, such cost escalation is a proper element of delay damages.

Labor cost escalation is typically incurred where the balance of the contract work following a significant delay must be performed at higher wage rates than would have been the case absent the delay. For example, where union labor is being used on a project, wage rates typically increase across the board as of a particular date, pursuant to the terms of a collective bargaining agreement. A compensable delay will therefore push some or all of the remaining contract work into the higher wage period, whereas it otherwise would have been performed before the wage increase went into effect. The compensable labor escalation cost in such a situation is measured by the wage increment (the amount of the raise) multiplied by the number of man-hours that were pushed into the higher wage period as a result of the compensable delay. Where multiple trades are involved, the effective dates of the wage increases and the amounts of those increases may well vary from trade to trade, making it necessary to calculate escalation costs on a trade-by-trade basis.

Calculation of labor escalation is often considerably more complex where non-union labor is involved, because wage increases may occur at different times and in different amounts for each individual worker. The records of the contractor may not even be specific enough to track precisely when each worker received an increase, or its amount. In such cases, an estimating procedure must typically be used to quantify the applicable labor cost escalation.

Material cost escalation is generally determined in the same general manner as labor escalation, by comparing anticipated material costs based on the originally planned schedule with the actual material costs incurred. This, however, creates an additional complication. A construction delay does not always mean that materials were necessarily or even properly purchased later than planned. This is particularly the case with respect to major items of permanent equipment (like transformers, motor control centers, boilers, chillers and the like), which are often long-lead-time purchases that would not normally be postponed due to a construction delay. Other construction materials, such as rebar or concrete, are often purchased only when needed, under contracts that provide for price increases on particular dates, making the cost escalation calculation relatively easy.

Material cost escalation is best demonstrated through invoices and receipts from which the actually experienced material price increases can be determined. If invoices or other actual cost records are not available, published indices of material price increases are frequently used instead, although this method is not preferred, since it relies on averages for the region rather than data specific to the project.

B. Liquidated Delay Damages

The technique of liquidating damages in advance by the terms of the contract is often used in construction contracts, most frequently to establish the delay damages to which the owner will be entitled should the contractor not complete the work by the contract completion date. The liquidated damages for delay are generally stated as a rate per day that the contractor will be liable to pay if the construction is not completed on time. Liquidated damages rates can vary over an extremely wide range, from as little as $100 per day for small projects, to over $100,000 per day on very large projects.

Liquidated damages provisions will be enforced in accordance with their terms so long as they are not a penalty, consistent with the general rule that contractual penalties are unenforceable. Accordingly, to be enforceable, a liquidated damages amount must be intended as liquidated damages and not as a penalty; it also must be based on a reasonable forecast of the actual damages that the owner would incur should the work not be completed as scheduled, determined as of the time of contracting. This last qualification is important, since in many cases at the time of the breach it is apparent that the liquidated damages rate is too high or too low in relation to the owner's actual damages. The owner in fact may have no actual damages at all stemming from the delayed completion. Nevertheless, the liquidated damages rate will be applied as agreed if the rate was reasonable as of the date of contracting.

C. Owner's Actual Delay Damages

In the absence of a liquidated delay damages provision, the owner is entitled to recover its actual damages due to the contractor's unexcused delay in completing the work as contracted. What these actual damages will consist of can vary as widely as the nature of the constructed facility varies. The damages attributable to delayed completion of a commercial building, such as a hotel, will differ dramatically from those stemming from delayed completion of a public work, such as a highway bridge. Nevertheless, some general categories of frequently encountered owner damages can be identified, including additional construction administration and supervision, lost rents and revenue, lost profits, loss of use and additional construction period interest.

1. Extended Construction Administration

In almost every instance of delayed completion, the owner will incur some direct damages, comparable to the contractor's extended site overhead costs, for continuing to oversee the contractor's work and administer the construction contract for an extended duration. Such damages include additional amounts paid to the architect, where the architect is providing construction administration services. The amount of these damages, however, is generally less than the contractor's site overhead because the owner's/architect's oversight role during construction is limited compared to the contractor's responsibilities.

2. Lost Rents, Revenues, Profits and Loss of Use

For projects that will generate a revenue stream (such as rents from tenants in an office building or shopping center) once they are completed and in operation, the bulk of the owner's actual delay damages usually take the form of lost rents, lost revenues of another sort or lost profits from the enterprise that was to occupy the completed project. The amount recoverable as damages of this nature will be largely dependent on the economics of the particular project, and what specific losses can be demonstrated to be the direct result of the delay experienced in project completion. Some typical key issues that are presented in damages claims in the nature of lost revenues or profits relate to:

a. How definite or speculative is the claimed loss? Rents on an unleased speculative office building or strip shopping center, or profits from a new business, are considerably more speculative in nature than rents from signed leases or revenues from an established business.

b. How foreseeable were the claimed losses? Lost rents, revenues or profits are generally characterized as consequential damages, which are recoverable only to the extent reasonably foreseeable at the time of contracting. If the contractor had no reason to know about the specific business deals on which the owner's claimed damages are premised, those damages may be beyond what was reasonably foreseen.

c. Is recovery of these types of consequential damages limited or precluded by contract? Many construction contracts contain consequential damages waivers or exclusions; in such cases, these waivers or exclusions often specifically address lost profits or revenues, lost customers or sales and the like. In most states, such provisions will be given effect, if unambiguous.

Another variant on this type of damages are damages for loss of use of the completed project. For example, the damages for delay in completion of a new facility being built by the owner for its own use, replacing a rented facility or one that has significantly higher occupancy costs, would not be the loss of revenue from the new facility, but the additional costs expended as a result of not being able to move into the new facility when planned. A claim for this type of damages, however, raises similar issues of foreseeability and possible preclusion by a consequential damages waiver as a claim for loss of revenue.

3. Additional Construction Period Interest

When the owner has obtained a construction loan to finance the construction project, another element of the owner's damages for delayed completion is likely to be the additional interest paid on that loan until the project reached completion. In many cases, the issues presented by this type of claim are subtle, however, and require a thorough understanding of the owner's financing arrangements. For example, the floating rate construction loan obtained by the owner may well roll over into permanent financing at a fixed interest rate. The real effects of the delay, therefore, may only have been the difference in interest rate between the construction and permanent financing. Moreover, if interest rates were changing during the delay period, the damages may include an increased interest rate on the permanent financing. Alternatively, that rate could have decreased, benefiting and not damaging the owner.

D. No Damages for Delay Clauses

While not included in the most frequently-used form conditions of contract, clauses that provide for an extension of time as the sole remedy for the contractor in the event of construction delays, and precluding the recovery of monetary damages for such delays, are found in a significant number of construction contracts. Such "no damages for delay" clauses are nominally enforceable, but since they are a form of exculpatory provision that attempts to relieve the owner from monetary responsibility for its actions, as a rule they are narrowly construed and not applied beyond their specific terms.

In addition, the enforcement of a "no damages for delay" clause is generally subject to several well-recognized exceptions, generally including (the specific wording varies somewhat from state to state): (1) delays so unreasonable in duration as to constitute an abandonment of the contract; (2) delays caused by the owner's active interference with the contractor's performance; (3) delays beyond the contemplation of the parties; and (4) delays due to bad faith or willful actions. The liberality with which these generally-recognized exceptions are applied, however, varies considerably among the states. In some jurisdictions, it seems as if the exceptions virtually swallow the general rule of enforceability, while in others, the clause will be enforced except in egregious circumstances. Overall, the propriety of using and seeking to enforce a "no damages for delay" clause remains one of the more controversial topics in construction law.

III. DAMAGES FOR INEFFICIENCY AND ACCELERATION

A. Identifying Labor Disruption and Inefficiency

Loss of labor efficiency or "disruption" experienced by a contractor or subcontractor is another frequent subject of construction claims. The term "disruption" generically refers to situations where the contractor has expended additional effort (generally, labor man-hours) to accomplish the contract work, over and above what would have been needed in the absence of the disrupting event. Disruption or labor inefficiency often accompanies a significant construction delay, although disruption can also occur even in the absence of any delay.

Inefficiency on a construction project can take many different forms, depending upon the type of project and nature of the work affected. Because efficiency is intangible and therefore difficult to measure, proving damages for inefficiency or disruption can be extremely difficult. Even under the best of circumstances, inefficiency typically cannot be proved exactly. Contractors rarely, if ever, keep cost records that are sufficiently detailed to enable isolating that portion of the contractor's labor costs that were incurred due to compensable inefficiency. A reasonable estimate that satisfies the test of "reasonable certainty" is all that is required, however, once entitlement is established. Even so, proving inefficiency damages is still a very challenging task.

B. Methods for Measuring Inefficiency Damages

Inefficiency or loss of productivity can be calculated in a number of ways, and no single approach will fit all situations. Actual cost records should be used to the extent they are available, but estimates and expert testimony to support the reasonableness of the estimates are almost always needed. The expert's testimony should also be corroborated by contemporaneous project records and fact witnesses familiar with events during construction. Generally, courts will view with skepticism estimates of inefficiency that are strictly theoretical in nature and not firmly grounded in factual testimony or supporting records.

The possible methods for demonstrating inefficiency damages include direct measurement, percentage estimates, comparisons with unaffected work, loss of learning curve calculations, and the total cost and modified total cost methods. In theory, direct measurement of inefficiency should be possible in some instances, by using techniques such as time-lapse photography, videotape, or web cameras. In reality, however, this method has rarely (to date, at least) been utilized.

More commonly, a percentage estimate of the efficiency reduction experienced is applied, most typically utilizing some form of industry study of loss of productivity caused by prototypical situations (such as stacking of trades or excessive overtime). While sometimes used successfully, this is not a preferred method; at a minimum, corroborative evidence will generally be required to demonstrate that the estimates bear a close relationship to the actual loss of efficiency.

Generally, the soundest, reasonably available method for proving inefficiency damages is to compare the contractor's productivity on work performed during an unaffected or "normal" period, with productivity on the same work during the disrupted period (the "measured mile" approach). Clearly, this requires two meaningful periods, one affected and one unaffected, where comparable work was ongoing. Disputes regarding this technique tend to focus on the comparability of the work in the two different periods, as well as other factors that differ between the two periods chosen.

In some cases involving repetitive activities, learning curve theory can be utilized to demonstrate the loss of efficiency of the work force. Learning curve theory seeks to measure the well-proven reduction in labor effort required to complete a particular task as the task is repeated over and over by the same worker or crew of workers. A learning curve analysis generally must be supported by expert testimony that the learning curves relied upon are accurate, and that the use of learning curves is appropriate to the situation.

C. The Total Cost and Modified Total Cost Methods

The simplest, and, as a result, most commonly used method of attempting to prove labor inefficiency is some variant on the total cost or modified total cost methods. At its essence, the total cost method involves subtracting bid or expected costs from actual costs, and claiming the difference as the damages. While generally disfavored, the total cost method or some variation on it can be successful in some circumstances, with appropriate precautions.

Courts are traditionally reluctant to award damages based on the total cost method because it assumes a cause and effect connection between the claimed damages and the claimed breach of contract, rather than proving such a connection. Specifically, damages presented using the total cost method contain three underlying assumptions: (1) all of the claimed costs in excess of the original estimate are the result of the events that are the subject of the claim; (2) the original cost estimate was an accurate indication of the contractor's cost to perform, absent those compensable events; and (3) the contractor was efficient in performing the contract and did not cause any of the cost overrun itself.

Because of these assumptions and the consequent risk of over-recovery, there is a four-part test generally applied to determine whether use of the total cost method is appropriate:

- The nature of the work makes it impossible or highly impracticable to determine the damages with a reasonable degree of accuracy by any other method;
- The contractor realistically estimated the work;
- The actual costs incurred are reasonable; and
- The contractor is not responsible for the additional costs and expenses.

The key to successfully using or defending against a total cost approach in an inefficiency claim is to either support or challenge the underlying assumptions. Challenges to total cost claims can be based on the accuracy of the estimated costs, the reasonableness of the actually incurred costs, or on the contention that a contractor-caused problem contributed to the cost overrun. Similarly, an otherwise dubious total cost claim can be improved by modifying the total cost claim, such as by using a "corrected" estimate that corrects for bid errors, or by excluding questionable actual costs, including those which were caused by the contractor or its subcontractor.

D. Acceleration Costs

Acceleration occurs where the contractor or subcontractor attempts to complete the project in less time than planned, due to actions attributable to the other party to the contract. There are two different types of compensable acceleration, express or ordered acceleration and constructive acceleration. Constructive acceleration requires a combination of events that have the same net effect as a directive to accelerate, including:

- An excusable delay that entitled the contractor to a time extension;
- Denial of the requested extension, coupled with
- A demand to complete the work in accordance with the original schedule; and
- Efforts to actually accelerate completion, whether or not successful.

Typically, recoverable acceleration damages include premiums paid for overtime labor, loss of efficiency associated with acceleration efforts, and costs for added equipment and materials needed to support an accelerated schedule. Adding multiple shifts and working Saturdays and Sundays are typical responses when attempting to speed completion, and these steps will usually generate added labor hourly costs, and often cause a loss of labor efficiency. Even when more craft labor is available and can be added in lieu of working longer hours, simply adding more tradespeople to a limited work space often results in a decrease in efficiency, even as the pace of work is increased.

In some cases, the contractor will use additional equipment or temporary materials as a means of accelerating the work, such as by bringing in an additional crane to speed a building project or additional earth moving equipment to accelerate a grading project. Alternatively, a contractor may increase its use of consumables or temporary materials. A typical example here is concrete forms. In order to accelerate concrete placement, the contractor often builds or rents more reusable concrete forms so that different areas can be poured simultaneously rather than sequentially. The rental cost of the added forms, or the cost of the lumber and labor to construct them, would not have been incurred but for the effort to accelerate the work, since without the acceleration, the existing forms would simply have been reused more times.

IV. DAMAGES FOR DEFECTIVE CONSTRUCTION

A. Cost to Repair

When construction work is incorrectly installed, or otherwise proves deficient in workmanship or materials, the basic measure of damages for the defective work is the cost of repair. Specifically, the cost of repair encompasses the reasonable cost to restore or repair the defective work to the condition required to conform to the contract specifications. This rule is potentially tempered in those situations where the owner has had the use of the work for some period before the defect appeared. In that case, some credit against the cost of repair may be due for that portion of the work's useful life that was obtained by the owner before the problem was encountered.

B. Diminution in Value

In some situations, the cost to repair defective work would be extremely high compared to the seriousness of the defect. The classic example is water piping installed in a house, covered up by the drywall and finishes, that is of a lesser grade than that which was specified. If the lower grade pipe that was installed is adequate for the intended purpose, though technically deficient because it does not meet the specifications, tearing the piping out of the walls to replace it with the proper grade (and destroying a good deal of the drywall and finishes along the way) would be a disproportionate remedy to fix a relatively minor deficiency.

Accordingly, the common law has evolved an alternative measure of damages to cover such situations. Where repairing the defect would require unreasonable economic waste, the cost of repair is not used as the measure of damages; instead, the damages are limited to the diminution in value caused by the defect. Under the diminution in value rule, the damages for the non-specification water pipe installed in a house would be limited to the difference in the value of the house (as typically determined by its market price) resulting from the substitution of a lesser grade of pipe. Almost invariably, use of this alternative measure of damages reduces the recoverable damages dramatically; accordingly, as a practical matter, the diminution in value measure has remained quite limited in its application.

V. DAMAGES FOR DEFECTIVE DESIGN

Damages for defective design include both the cost to correct the defective design documents themselves, and the cost to correct work that has been installed in accordance with the defective design. When it is the design or engineering that is defective, however, instead of the construction, there are several additional issues that arise with respect to determining the proper measure of damages.

One such issue relates to the theory of recovery. In most cases, a claim for defective design is necessarily premised on professional negligence, a failure to conform to the applicable standard of care for architects or engineers with respect to a particular aspect of the project's design. When asserted by the party having a contract with the design professional (usually the owner), this claim is phrased in terms of negligence—a tort concept—but (as most states recognize) is nevertheless still based on the contract. When asserted by injured third parties (as discussed in the next section), the claim may be based on contract (under a third-party beneficiary theory), but is more often a true tort claim for negligence. This confusing situation spawns equal confusion respecting whether the measure of damages is determined by contract law principles or the somewhat more liberal tort law standard of joint and several liability for damages proximately caused by the negligent act.

A second such issue relates to the fact that defective design can take the form either of a design error, where what was specified and installed is incorrect and needs to be modified or corrected, or of a design omission, where a necessary element of the design was simply not included and has to be added after the problem is discovered. The effect of this distinction on the recoverable damages is explored further below.

A. Cost to Correct Erroneous Design Documents

The most obvious element of damages for defective design is the cost to correct the defective design documents themselves. Architect/engineer contracts frequently provide that design documents containing errors or omissions will be corrected at no additional cost to the owner. Those same contracts, however, sometimes also limit the owner's remedy to correcting the documents. Such a limitation often provides almost no effective remedy at all. If the work has been installed, correcting the actual field construction done in accordance with flawed drawings and specifications is almost certain to involve far greater cost than correcting the design documents themselves.

B. Cost to Correct Design Errors and Omissions and the Concept of Betterment

In the case of clear design errors, the recoverable damages include the reasonable cost to correct or repair the error.

The correction costs are limited, however, to those required to bring the design within the appropriate design standard, and not those to alter the design in accordance with the owner's preference, which may be well in excess of what would meet the professional standard of care.

In the case of design omissions, the assessment of damages begins, but does not end, with the reasonable cost to add the omitted, but necessary, item or design feature. In most cases, the omitted item, had it been included in the original design as it should have been, would have increased the overall cost of the project. In other words, the owner would have paid more to the contractor from the outset to construct the properly designed project than it actually did. In this sense, remedying the omission adds value to the project and to that extent represents betterment. The betterment element is accordingly not a damage caused by the defective design.

The dilemma this presents in design omission cases is how to identify and quantify the betterment element, and deduct it from the total cost to add the omitted item or feature. If the omission is discovered only well after construction is underway, and corrected by issuing a change order to the contractor, the cost of remedying the omission in this manner is arguably more than if the omitted item had been included in the original design and encompassed by the contractor's original bid. The same concept holds even more true when the omission is corrected only after construction is completed. The damages from the design omission in such situations, however, is limited to the incremental cost of remedying the omission at a later date, to the extent that cost is over and above what would have been paid to include the omitted item in the original design.

VI. DAMAGES FOR THIRD-PARTY CLAIMS

Third-party claims, for purposes of this discussion, refer simply to claims by persons with whom the party against whom the claim is asserted does not have a contract. Typical third-party claimants in construction cases include tenants and other ultimate users of the project, injured construction workers and passersby, and parties to the construction process who are not in privity of contract with the party against whom the claim is asserted, such as contractor or subcontractor claims against the architect. Third-party claims may be based on a third-party beneficiary theory, in which case they remain governed by the law of contract damages. More typically, however, they are based on a tort theory such as negligence or trespass, in which case the recovery of damages for them may well be limited by the economic loss rule.

A. Property Damage and Physical Injury

Tort law classically is concerned with injuries to persons or physical damage to property, such as that caused by a collapse or other physical failure. With otherwise viable third-party claims sounding in tort, damages of this nature are generally recoverable. This includes the economic components, such as medical bills, lost wages and the cost to rebuild physical damage to property, as well as non-economic damages, such as pain and suffering.

B. The Economic Loss Rule

The economic loss rule has developed in response to the proliferation in recent years of efforts to use tort theories to recover damages beyond the traditional tort concerns of physical injuries to persons or tangible property. The economic loss rule holds that recovery of purely economic losses is properly the realm of contract law and not tort law, as tort law was never intended to protect harm to economic interests. As such, under the economic loss rule, damages for purely economic losses are not recoverable in tort. About 40 states today enforce some form of economic loss rule, although the strictness of the rule varies significantly.

The problem arising from application of the economic loss rule in the construction context is that the damages deemed unrecoverable as "pure economic losses" include most damages sought in most construction claims. For example, the damages claimed by a contractor against an architect for negligent design errors, or unreasonable delays in issuing design documents or returning shop drawings, are generally considered purely economic losses incurred by the contractor, and thus unrecoverable in tort. Where a negligent defect in the work is discovered in advance of a physical failure, and costs are incurred to prevent the failure and the consequent physical damage to property, these costs are nevertheless still considered unrecoverable economic losses under the stricter versions of the economic loss rule.

Similarly, property damage to the defective component itself, as opposed to other property damaged by a physical failure of the component, may also be considered unrecoverable economic loss. In sum, where it is applicable, the economic loss rule works a substantial limitation on the availability of tort theories to recover construction damages.

VII. OTHER CONSTRUCTION DAMAGES

A. Punitive Damages

Punitive damages are damages awarded not to compensate for a loss, but to punish and to serve as an example to others. Punitive damages are generally not recoverable as a matter of law on claims for breach of contract, because breach of contract is considered an economically driven decision, and there is no component of moral wrongdoing associated with breaching a contract. Punitive damages are therefore usually reserved for tort actions involving gross negligence, willful or malicious conduct, fraud, or breach of a fiduciary duty—all situations only rarely present in construction claims.

Under the laws of most states, there must be an independent tort to recover punitive damages and not just a willful or egregious breach of the contract. In a minority of states, however, parties to a contract are considered to be subject to an independent duty of good faith and fair dealing, and an egregious breach of that independent obligation can serve as the basis for an award of punitive damages.

B. Other Types of Liquidated Damages

While liquidated damages clauses are most frequently used to establish the owner's recoverable delay damages, other types of damages that will result from particular contract breaches may also be liquidated in advance by contract. For example, design-build contracts to construct power plants and other industrial plants frequently include guarantees of the performance of the completed plant. Such contracts also commonly include a form of liquidated damages clause, known as a "buy-down" clause, that specifies the damages payable for a specific degree of shortfall from each guaranteed performance parameter, with the damages increasing with the degree that actual performance falls short of the guaranteed level. Such clauses are enforceable consistent with the enforceability of liquidated damages clauses generally.

Another example of an alternative form of liquidated damages clause is often seen on very critical, time sensitive repair or renovation projects. For example, a contract to repair a bridge may provide that a failure to open all lanes of the bridge to traffic by 6:00 a.m. on workdays will result in liquidated damages of $5,000 or more for every 10 minute period the lanes are not open. Even though the public agency imposing those high liquidated damages cannot show that its added costs even approach a figure like $30,000 per hour, such clauses have been upheld on the basis that the public agency can properly base the liquidated damages amount on an assessment of the inconvenience and lost time to the commuters using the bridge generally.

As these examples indicate, liquidated damages clauses can be creatively and effectively utilized in a wide variety of construction situations. They are not just limited to damages for delays in completion.

C. Profit

Given the relatively tight restrictions on the recoverability of contract damages in other respects, it is perhaps surprising that recovery of a percentage markup for profit is commonly an allowable element of construction damages in claims by the contractor. Most courts will allow a reasonable markup for profit in a breach of contract case, on the theory that the breaching party should not be allowed to utilize the resources of the contractor, and deprive it from performing profitable work for others, without paying the contractor a reasonable profit for its efforts required by the breach. The same theory, however, does not apply to the owner, who is paying for construction services and not expecting any profit from the construction process. Hence, profit is not generally allowed on the owner's damages (except possibly in the form of lost profits on the completed project, as previously discussed).

Construction contracts frequently contain provisions that limit the allowable percentage markup for profit, as well as overhead, and these will be enforced where applicable. In any event, the amount of profit must be reasonable to be recoverable. Often, the profit markup in the contractor's original bid is utilized, if reasonable, and a 10% profit markup is also frequently seen.

D. Interest

The availability and rate of prejudgment interest is a matter of state law, and is increasingly addressed by statute rather than left to the common law. The general common law rule in most states is that prejudgment interest is recoverable on contract damages only when the damages are considered as liquidated.

The standard for determining when contract damages are considered as "liquidated," however, varies considerably from state to state. At the liberal end, damages are considered liquidated if they are theoretically *capable* of being ascertained specifically, even though they may be in considerable dispute until the conclusion of a protracted litigation.

More conservative interpretations require the damages to be reasonably ascertainable under a contract provision or by some other reasonably objective means before they are considered "liquidated." State law also tends to vary significantly as to when prejudgment interest starts, and the allowable interest rates range from 5% to 12% or more.

Because of these substantial variations and the proliferation of statutes generally liberalizing the recovery of interest in contract cases in recent years, determination of the state's rules on the recovery of prejudgment interest should be checked at an early point in a dispute. Additionally, if the relevant state law on recovery of prejudgment interest is unfavorable, it may be possible in some circumstances to recover as an element of damages the interest actually paid on a loan obtained specifically to finance the expenditures resulting from the breach. While this rule is of limited application, it can sometimes provide a method to recover interest expenditures.

Recoverability of interest on Federal government contract claims is governed by the Contract Disputes Act of 1978. That act provides for recovery of simple interest on a claim at a Treasury Department-prescribed rate from the date the certified claim is received by the Contracting Officer until the date of payment.

E. Attorneys' Fees and Litigation Costs

The "American rule" on attorneys' fees, generally followed in the United States, means that each party normally pays its own legal fees and most associated costs, whether it wins or loses in litigation. The "costs" that are awarded to the prevailing party in most courts or in arbitration actually represent only a small fraction of the typical litigation expenses. Construction contracts will on occasion contain a provision that the prevailing party in a dispute can recover its attorneys' fees and litigation costs. Such provisions are enforceable and will be given effect; if vaguely worded, however, they may be interpreted narrowly by the courts.

In rarer instances, a state statute may specifically provide for recovery of legal fees by the prevailing party, but this is not the general rule in contract actions in any state. On Federal government contracts, the Equal Access to Justice Act (EAJA) permits small businesses that are successful in actions against the Federal government to recover their litigation costs, including statutory attorneys' fees.

VIII. CONCLUSION

Gaining a mastery of the issues involving construction damages requires not only knowledge of the applicable legal principles, but also considerable care and attention to detail. Many disputes are won or lost on the basis of the care and accuracy with which the damages have been recorded, documented, summarized, supported and presented. Demonstrating the necessary cause and effect relationship between the compensable event and the claimed damages can be tedious; assembling and presenting the detailed oral and documentary evidence to substantiate the reasonableness of the claimed damages can be time-consuming and expensive. The reward, however, is both tangible and proportionate to the toil, just as the risks of giving damages issues short shrift are many and perilous.

Construction Insurance: An Introduction

12

James P. Wagner

Sooner or later, every construction practitioner faces issues or questions relating to insurance. Whether preparing and negotiating a construction agreement or litigating a claim for damages for defects or deficiencies arising out of construction, insurance is often an important consideration. Thus, construction practitioners should have a basic understanding of the coverages provided by the policies typically offered by insurers to owners, contractors, and design professionals.[1]

Most property and casualty insurers utilize standard policy forms developed by Insurance Services Office, Inc., or "ISO," a firm which provides a wide variety of information and services to the insurance industry. Accordingly, most references herein to "typical" or "standard" policy provisions refer to terms appearing in standard ISO policy forms.

---

1. For a detailed review of construction insurance coverage, consult DEUTSCH, KERRIGAN & STILES, CONSTRUCTION INDUSTRY INSURANCE HANDBOOK (1991).

I. INSURANCE COVERAGES MANDATED IN STANDARD CONSTRUCTION CONTRACTS

Most construction contracts require the purchase and maintenance of insurance by some or all of the parties. Indeed, prudence dictates that the parties to the construction process (like other businesses) maintain insurance to cover liability exposures that can occur. There are few industries that experience the risk of loss that the construction industry experiences. Most construction projects require sizeable commitments of labor, equipment and materials. Many separate entities come together over a period of time to produce completed facilities. In these circumstances, the risk of an occurrence resulting in damage to the project or to some third party is significant. Thus, from the project owner's perspective, it is important that those with whom it contracts maintain appropriate insurance to cover such risks.

Typically, proof of insurance coverage is provided by a "Certificate of Insurance," a standard form which an insured can obtain from his or her insurance broker describing the nature and amount of insurance coverages maintained by the insured, as well as any entities holding status as an "additional insured" under those policies. A Certificate of Insurance, however, is issued only for informational purposes; it is not an actual policy of insurance itself. The scope of coverage actually provided can only be determined from the policies themselves.

Standard form contracts such as the AIA A-201 ("A201") set forth the contracting parties' obligations with respect to the purchase of insurance. Under § 11.1.1 of the A201, a contractor is required to purchase and maintain insurance that will cover:

- claims under the workers' or workmen's compensation, disability benefit and other similar employee benefit acts which are applicable to the work to be performed;
- claims for damages because of bodily injury, occupational sickness or disease or death of the contractor's employees;

- claims for damages because of bodily injury, sickness or disease, or death of any person other than the contractor's employees;
- claims for damages insured by usual personal injury liability coverage which are sustained by a person as a result of an offense directly or indirectly related to employment of such person by the contractor, or by another person;
- claims for damages, other than to the work itself, because of injury to or destruction of tangible property, including of loss of use resulting therefrom;
- claims for damages because of bodily injury, death of a person, or property damage arising out of ownership, maintenance or use of a motor vehicle; and
- claims involving contractual liability insurance applicable to the contractor's obligations under paragraph 3.18 (providing for indemnity).[2]

Most of the coverages required under AIA Document A201 are provided under policies known as commercial general liability ("CGL") policies. The other required coverages are generally provided under worker's compensation and motor vehicle insurance.

Under the A201 (§ 11.4), the owner is required to provide property insurance for the project. The coverage is required to be on an "all-risk policy form," and must insure against "fire (with extended coverage) and physical loss or damage including, without duplication of coverage, theft, vandalism, malicious mischief, collapse, earthquake, flood, windstorm, falsework, testing and start up, temporary buildings and debris removal including demolition occasioned by enforcement of any applicable legal requirements."[3] This coverage is generally provided by a "builder's risk" policy the owner

2. AIA Document A201 § 11.1.1 (1997 ed.).
3. AIA Document A201 § 11.4.1.1 (1997 ed.).

obtains. According to one commentator, since the owner generally pays the cost of builder's risk insurance no matter who procures the policy, "it may as well have the control and security that goes with buying insurance, because property loss and insurance disputes can slow a job, imperiling the owner's position, even if the owner itself lost no property."[4]

Many owners, developers and general contractors include a provision in lower-tier contracts requiring that they be named as an "additional insured" on liability policies carried by their subcontractors. The nature and significance of "additional insured" status will be discussed below. Where such a requirement is set forth, the lower-tier contractor must make its insurance broker aware of this requirement and make certain that the insurer actually issues the necessary endorsement to avoid possible liability for breaching this requirement.

With respect to design contracts, while the standard AIA contract between the architect and the owner (Document B141) does not specifically require the architect to maintain insurance, the document recognizes that the design professional likely will carry insurance coverage, including professional liability insurance.[5] Consistent with this recognition, most design professionals carry both CGL coverage and professional liability insurance. If the client contemplates utilizing a non-standard form of contract, counsel should verify that the form utilized provides that all appropriate insurance coverages will be obtained.

A. Commercial General Liability Insurance

Understanding the scope of coverage afforded by CGL insurance policies is essentially a two-step process. One must first understand the broad scope of coverage afforded

---

4. DEUTSCH, KERRIGAN & STILES, *supra* note 1, § 9.1 at 144.
5. AIA Document B141 § 1.3.9.2.6 (1997 ed.).

by the "insuring agreement," and then address the various "exclusions" which purport to narrow that coverage.

1. The Insuring Agreement

a. *"Bodily injury" or "Property damage"*

CGL policies provide very broad coverage for damages arising from "bodily injury" or "property damage." While policy forms may differ somewhat between companies, they typically provide, under "Coverage A—Bodily Injury and Property Damage Liability," that the insurer will agree to pay those sums the insured is *legally obligated to pay as damages because of bodily injury or property damage caused by an occurrence.*[6]

Note that this grant of coverage does not limit coverage for "property damage" to "property of others" or otherwise. The coverage grant is broad. Most limitations on coverage are found elsewhere in the policy, primarily in the Exclusions section. This is significant because, in the event of a dispute, the insured's burden of proof extends *only* to establishing that the loss falls within the coverage grant. It is the *insurer* that has the burden of proving the application of one or more policy exclusions.[7] The insured's burden of proof is simply to establish the existence of a claim for "bodily injury" or "property damage" caused by an "occurrence."[8]

The liability theory of the underlying claimant is not dispositive with respect to coverage. Thus, even if a claim is labeled "Breach of Contract," this should be irrelevant for

---

6. Insurance Services Office, Inc. ("ISO") has produced a standard CGL policy after which most CGL policies are patterned. The CGL policy provisions discussed herein are based upon the 1986 edition of the ISO CGL insurance form.

7. *See, e.g.*, Fireman's Fund Ins. Co. v. Ex-Cell-O Corp., 750 F. Supp. 1340, 1356 (E.D. Mich. 1990); Brown v. Snohomish County Physicians Corp., 102 Wn.2d 747, 845 P.2d 1334, 1341 (1993). *See generally* 2 WINDT, INSURANCE CLAIMS AND DISPUTES § 9.01.

8. Since this paper concerns construction defect claims, we focus our attention on claims for "property damage."

purposes of coverage. Instead, the correct analysis is whether the claim is one for "property damage." The definition of "property damage" in most CGL coverage forms includes some version of the following:

1. Physical injury to tangible property, including all resulting loss of use of that property. All such loss of use shall be deemed to occur at the time of the physical injury that caused it; or
2. Loss of use of tangible property that is not physically impaired. All such loss of use shall be deemed to occur at the time of the "occurrence" that caused it.

The key concepts within this definition are "*physical injury*"—the property must have sustained some type of physical harm or damage; and "*tangible property*," which typically precludes coverage for *purely* economic losses such as lost profits or delay damages. Generally, purely economic losses do not constitute damage to "tangible property."[9] Note, however, that if such economic losses occur *because of* property damage, the losses may be covered. The determining factor is the presence of property damage, beyond the cost of completing the work or replacing any defective work.

Under paragraph (b) quoted above, loss of use of tangible property is covered under the CGL form, regardless of whether that property suffers physical injury. Thus, for example, in *Commercial Union Insurance Co. v. R.H. Barto Co.*,[10] where the insured installed a defective HVAC system and the building was rendered unusable for a period of time as a result, the claim was covered as a "loss of use," even though there was no "physical injury" at all. Thus, defective

9. *See* Lamar Truck Plaza, Inc. v. Sentry Ins., 757 P.2d 1142 (Colo. 1988) ("purely economic loss" does not constitute loss of use of tangible property within definition of property damage); Sentry Ins. Co. v. S&L Home Heating Co., 414 N.E.2d 1218 (Ill. Ct. App. 1980) ("The policy defines 'property damage' as 'injury to or destruction of tangible property.' The claimed loss of productivity . . . is economic loss and not property damage.").

10. 440 S.2d 383 (Fla. D. Ct. App. 1983), *rev. denied*, 451 S.2d 850 (Fla. 1984).

construction which results in a "loss of use" may well be covered even where no "physical injury" has occurred.

One issue that occasionally arises is whether diminution *of the value* of a structure, without other damage, constitutes "property damage." Some courts have held that diminution in value does not constitute "physical injury." For example, in *American Home Assurance Co. v. Libby-Owens-Ford Co.*,[11] the First Circuit stated that the 1973 change to the standard CGL policy, which expressly added the modifier "physical" injury to the definition of "property damage," precluded claims for intangible losses such as diminution of value.[12] Some commentators argue, however, that diminution in value is only the *measure* of damages, and that the physical injury is the physical presence of the unwanted defective work.[13] A few courts have permitted coverage for diminution in value cases, particularly when the work in question would be impossible or highly impractical to repair.[14]

b. ". . . Caused by an Occurrence."

The other important prerequisite to coverage under the insuring agreement is an "occurrence." Most policies define an "occurrence" as some version of the following:

> . . . an accident, including continuous or repeated exposure to substantially the same general harmful conditions.

Many "occurrences" are easily identified, *e.g.*, the collapse of a wall. The more difficult cases are those that involve "con-

---

11. 786 F.2d 22 (1st Cir. 1986).

12. *See also*, *New Hampshire Ins. Co. v. Viera*, 930 F.2d 696 (9th Cir. 1991). Remember, however, that the definition of "property damage" specifically allows for loss-of-use claims, regardless of any physical injury to tangible property.

13. *See, e.g.*, Kalis, et al., *Policyholders Guide to the Law of Insurance Coverage* § 7.01(b) (1997).

14. *See, e.g.*, *Missouri Terazzo Co. v. Iowa National Mut. Ins. Co.*, 749 F.2d 647 (8th Cir. 1984); *Eljer Mfg. v. Liberty Mutual Ins. Co.*, 972 F.2d 805 (7th Cir. 1992), *cert. denied*, 113 S. Ct. 1646 (1993).

tinuous or repeated exposure to substantially the same general harmful conditions." In some cases, carriers have argued that defective construction does not constitute an "occurrence." A few courts have accepted the companies' position, with one stating that "construction defects . . . are natural and ordinary consequences of improper construction techniques . . . and, thus, do not constitute an occurrence within the definition in the CGL policy."[15]

Most courts, however, reject such a narrow interpretation of the coverage grant, and hold that defective construction can result in an "occurrence," so long as the damage was "unexpected and unintended."[16] Where defects in construction result in "continuous or repeated exposure" to harmful conditions, coverage should be triggered. Water intrusion cases present classic examples of this type of "occurrence."

Notably, cases involving "continuous or repeated exposure to substantially the same general harmful conditions" frequently result in the triggering of more than one policy year's coverage. For example, in *Gruol Construction Co. v. Insurance Co. of North America*,[17] the builder of an apartment building was sued for damage to the building caused by dry rot which had resulted from dirt having been left piled against the box sills of the building during construction. The builder tendered defense to three insurance carriers who had provided coverage for six years following the completion of the work. All three refused to defend. The Washington State court held that all of the policies at issue provided coverage, since the "occurrence" was a continuous process which was ongoing throughout the several policy periods.

The question of when an "occurrence" triggers coverage can be extremely complicated; it has been the subject of multiyear, multimillion-dollar coverage wars involving cleanup of

---

15. *Monticello Ins. Co. v. Will-Fred's Construction*, 661 N.E.2d 451, 455 (Ill. App. 2d Dist. 1996).

16. *Havenstein v. Saint Paul-Mercury Ins. Co.*, 65 N.W.2d 122, 124 (Minn. 1954); *Maryland Casualty v. Reeder*, 270 Cal. Rptr. 719 (Cal. App. 1990).

17. 524 P.2d 427 (Wn. App. 1974).

Superfund sites. Under any formulation, however, the time of an "occurrence" is not the time when the defective work was performed, but the time the *property damage* occurred.[18]

Different courts view this issue differently. Some take the position that the "occurrence" occurs when the property was *first exposed* to the dangerous condition (the "exposure" theory). Other courts define the "occurrence" as having taken place when the damage "*manifested*" itself (the "manifestation" theory). Some states define the "occurrence" to incorporate *the entire time period from the first exposure to the "manifestation,"* and others define the "occurrence" as happening *at any time damage actually occurs* (the "injury-in-fact" theory).[19]

2. Exclusions

An understanding of the most commonly cited exclusions, and how they have been interpreted by the courts (not always consistently), is essential to understanding the scope of coverage afforded under CGL policies. Although a detailed discussion of the numerous CGL policy exclusions is beyond the scope of this chapter,[20] some of the more commonly encountered exclusions are discussed below.

a. Damage To Property Exclusions

(1) "Property you own or occupy" This exclusion denies coverage for damage to property that is either owned or occupied by the insured. In the construction defect context, this exclusion is often raised against claims by developers of resi-

---

18. 11 *Couch on Insurance Second* §§ 44:256-57 (1982).

19. *See Sentinal Ins. Co. v. First Insurance Co. of Hawaii*, for a discussion of the differing coverage "triggers," including a number of useful citations. 875 P.2d 894, 914-919 (Haw. 1994).

20. For a detailed analysis of CGL policy exclusions, see SCOTT C. TURNER, INSURANCE COVERAGE OF CONSTRUCTION DISPUTES §§ 8–40 (West 1999).

dential projects where the developer continues to own at least part of the property after construction is completed. However, since most contractors do not own or lease the allegedly damaged property, this exclusion is usually not an issue in claims against contractors. Even in cases involving owner-developers, this exclusion should only apply to that *portion* of the premises still owned by the insured.

(2) "Care, custody, or control" The term "care, custody or control" as used in an insurance policy has been defined as having the right to exercise dominion over the subject matter. The 1986 version of the CGL policy narrowed the scope of this exclusion by applying it only to personal property. As such, in the context of construction defects, the only time this exclusion may come up is where there is a question of whether the particular property is "personal" or "real."[21]

(3) "Business Risk" exclusions The so-called "business risk" ex-clusions are intended by insurers to bar coverage for "normal, frequent, or predictable consequences of doing business, and which business management can and should control and manage."[22] The most important point to remember about these exclusions is that they do not apply to any claims arising after the insured has completed its work.

(a) "Performing operations" This exclusion denies coverage for damage to "that particular part" of property on which the insured is performing its operations, if the damage arises out of those operations. It excludes:

> that particular part of real property on which you or any contractors or subcontractors working directly or indirectly on your behalf are performing operations, if the 'property damage' arises out of those operations.

---

21. *See National Union Fire Ins. Co. v. Structural Systems Technology, Inc.*, 964 F.2d 759 (8th Cir. 1992).

22. *Columbia Mutual Ins. Co. v. Schauf*, 968 S.W. 2d 74, 77 (Mo. 1998) (citations omitted).

This particular exclusion is limited to the property damaged while the insured was actually working on it.[23]

(b) "Faulty workmanship" This exclusion denies coverage for the cost to repair or replace an insured's own faulty or defective work. It states:

> . . . coverage is excluded for that particular part of any property that must be restored, repaired or replaced because your work was incorrectly performed on it.

Insurers believe this exclusion differentiates CGL insurance from a performance bond. This exclusion does not apply, however, to property damage *resulting* from faulty or defective work.

For instance, in *W.E. O'Neil Constr. Co. v. National Union*,[24] an owner of a garage sought damages from a contractor after the concrete used in the garage experienced excessive cracking and made the garage unusable above the first floor. The cause of the cracking was improper placement of steel mesh that was supposed to be imbedded in the concrete to ensure the structural integrity of a garage. The court held that the "faulty workmanship" exclusion excluded the cost of repairing or replacing the defective steel mesh; the court found coverage, however, for the resulting property damage to the concrete and other elements, even though it was caused by faulty workmanship.

In addition, this particular exclusion, unlike the "performing operations" exclusion discussed above, does not include the phrase "contractors or subcontractors working directly or indirectly on your behalf." Thus, this exclusion does not apply to work done on the insured's behalf by a contractor or subcontractor.[25]

---

23. *Fisher v. American Family Mutual Ins. Co.*, 579 N.W.2d 599, 604 (ND 1998); *Columbia Mutual Ins. Co. v. Schauf*, 967 S.W. 2d 74, 81 (1998).

24. 721 F. Supp. 984 (N.D. Ill. 1989).

25. *See Kalchthaler v. Keller Constr. Co.*, 591 N.W. 2d 169 (Wis. Ct. App. 1999); *Fireguard Sprinkler Systems, Inc. v. Scottsdale Ins. Co.*, 864 F.2d 648 (9th Cir. 1988).

Moreover, this exclusion does not apply to "property damage" included in the "products-completed operations hazard." "Products-completed operations hazard" is defined as:

> Products-completed operations hazard includes all "bodily injury" and "property damage" occurring away from premises you own or rent and arising out of "your product" or "your work" except:
>
> 1. Products that are still in your physical possession; or
> 2. Work that has not yet been completed or abandoned arising out of "your product" or "your work" except products that are still in your physical possession or work that has not yet been completed or abandoned.

In plain English, this means that the faulty workmanship exclusion does not apply if the property damage occurs after the insured has completed its work.[26]

(c) The "Products-Completed Operations Hazard" and the "Your Product" and "Your Work" Exclusions As stated above, the "products-completed operations hazard" covers claims, including most defective construction claims, which arise after construction has been completed on a project. The exclusions most commonly cited with respect to the "hazard" are those relating to "your product" and "your work."

Some carriers attempt to exclude claims arising out of defective construction by characterizing the completed project as a "product," and relying on the "your product" exclusion. Most courts, however, have rejected the argument that buildings or other construction projects are "products."[27] Further, the definition of "your product" under the 1986 version of the CGL policy includes: "[a]ny goods or products, **other than real property,** manufactured, sold, handled, distributed or disposed of by You." Thus, it would seem that a completed construction project could not be a "product."

---

26. *See Kalchthaler v. Keller Constr. Co.*, 591 N.W.2d 169 (Wis. Ct. App. 1999).
27. *See Maryland Casualty Co. v. Reeder*, 270 Cal. Rptr. 719 (Cal. App. 1990).

The exclusion differs from the "faulty workmanship" exclusion in that it relates to completed work, excluding:

> Property damage to **your work** arising out of it or any part of it and included in the products-completed operations hazard.

> This exclusion does not apply, however:

> . . . if the damaged work or the work out of which the damage arises was performed on your behalf by a subcontractor.

Similar to the "faulty workmanship" exclusion, the "your work" exclusion denies coverage for costs to correct the insured's own defective work, but does not preclude coverage for damages to property other than the insured's own work. Moreover, and significantly, this exclusion "does not apply if the damaged work or the work out of which the damage arises was performed on [the insured's] behalf by a subcontractor."[28] This means that damage caused by defective workmanship by a subcontractor, *even damage to the insured's own work*, is covered.

Finally, as this exclusion concerns "work," it has been held inapplicable at least by one court to a claim arising solely from the supply of defective materials.[29]

B. Builder's Risk Insurance

Builder's risk insurance, purchased by the owner under the scheme contemplated by the AIA documents, is a specialized form of property insurance for projects under construction.[30] Such projects are not covered under standard property policies. Builder's risk policies are designed to insure against the accidental destruction or damage to the uncompleted construction project or the materials stored on site. Builder's risk insurance is similar to standard property insur-

---

28. *Spears v. Smith*, 690 N.E.2d 557, 560 (Ohio Ct. App. 1996).

29. *See Economy Lumber Co. of Oakland, Inc. v. Insurance Company of North America*, 204 Cal. Rptr. 135 (Cal. Ct. App. 1984).

30. The standard ISO builder's risk coverage is set forth in form CP 00 20 06 95.

ance to the extent that both forms insure for damage to a loss of property, regardless of fault. One commentator explained why standard property insurance is not used for buildings under construction, as follows:

> First, structures under construction frequently represent risks inherently and substantially greater and more difficult to evaluate than the same structure would when complete. Second, construction projects involve a special problem in the changing value of the risk; it is valued at zero or near zero at commencement and increases throughout the policy life to the full completed value. Third, because of the hazard to the insurer of arson for profit, fire insurers have traditionally required an identified "insurable interest," that is, an insured could insure property only to the extent it had an interest in the property. But there are a variety of interests in a construction project. Rigid definition of the insurable interests in a building under construction is much more difficult, or perhaps impossible, because the insurable interests fluctuate from day to day.[31]

Most builder's risk policies only provide coverage while the construction is ongoing; once construction is "completed," builder's risk insurance terminates. Coverage is typically provided for the "completed value" of the insured structure and should include all permanent fixtures and interior improvements that will be part of the completed structure. Note that the construction contract price will typically *not* equate to the full "completed value" of the project.

The "covered property" under the standard builder's risk form also differs from that addressed by standard property coverage. The standard form covers the building under construction and its foundation, fixtures and machinery, equipment used to service the building, and the insured building materials and supplies used in construction.

Typical ISO builder's risk policies are issued with one of three "cause of loss" forms; the "basic" (or "named perils") form, the "broad" form or the "special" (or "open perils")

---

31. DEUTSCH, KERRIGAN & STILES, *supra* note 1, § 9.2 at 145.

form. The "open perils" form, which offers the broadest possible protection, is the most popular with insureds and their agents. Other special endorsements may be procured (often at considerable additional cost) to further broaden coverage. The prospective project developer and contractor should carefully review the scope of appropriate builder's risk coverage with their insurance agents.

C. Insurance for the Design Professional

Design professionals generally carry professional liability insurance, in addition to CGL coverage. While the standard form CGL policy does not include a "professional liability" exclusion, most insurers will add an endorsement excluding it. The typical endorsement states that the policy does not apply to injury caused by "the rendering or failure to render any professional service" listed in the endorsement. "Professional services", however, are generally not defined in the CGL policies, and this is an area which has spawned considerable litigation. It is important to note that in applying the "professional liability" exclusion, the courts look to the nature of the alleged negligent act, and not to the title or character of the actor. Thus, courts have held that ministerial acts by a professional are not within a "professional services' endorsement.[32]

While these endorsements excluding liability arising from "professional services" are generally upheld, there are instances where the claims against a professional contain allegations of breach of both professional duties and non-professional duties. For example, a lawyer may be alleged to be negligent in referring a client to an incompetent lawyer and alleged to have defamed the client in a non-professional setting as well. Thus, it is possible that an exclusion for breach of professional duties would not apply to all of the alleged grounds for liability, and a

---

32. *See Guaranty Nat. Ins. Co. v. North River Ins. Co.*, 909 F.2d 133, 135-136 (5th Cir. 1990); *Duke University v. St. Paul Fire & Marine Ins. Co.*, 386 S.E.2d 762, (N.C. App. 1990).; *See also* Windt, *Insurance Claims & Disputes*, 3d ed., § 11.16 (1995).

claim could be within the scope of both a professional liability policy and the CGL policy.[33]

Generally, however, professional liability insurance policies are necessary to provide defense or indemnity for design professionals. Such policies are generally sold on a "claims made" basis. A "claims made" policy provides coverage for claims that are "first made" against the insured and reported to the underwriter during the policy period, even if the alleged breach of professional duty and injury may have occurred before the policy began.

A claims made policy is triggered by the existence of a "claim." Though the term is pivotal to the attachment of coverage, not all policies contain a definition of a "claim." One policy defined a "claim" as "a demand for money or services or the service of a suit or institution of arbitration proceedings."[34] A "claim" requires an assertion of a right for damages.[35]

Claims made policies generally have a provision referring to the insured's right to give notice to the insurer of facts indicating the potential for a claim, even though a "claim" has not yet been made. The giving of such notice will trigger coverage under the policy if a "claim" is ever made, *even if after the expiration of the policy.* The courts have relied upon this distinction in the policy to hold that an insured's knowledge of facts that indicate that a "claim" may eventually be made is not the equivalent of a "claim."[36] Even if the policy does not have a specific provision granting coverage for occurrences reported to the carrier during the policy period, courts have held that

---

33. *See, e.g., Guaranty Nat. Ins. Co. v. North River Ins. Co.*, 909 F.2d 133 (5th Cir. 1990) (hospital sued for patient's suicide; both professional liability and CGL policies applied because hospital had both professional and non-professional duties to patient).

34. *Tewell, Thorpe & Findley Inc. v. Continental Cas. Co.*, 64 Wn. App. 571, 574 n.1, 825 P.2d 724 (1992).

35. *James F. O'Connell & Assoc. v. Transamerica Indem. Co.*, 61 Wn. App. 103, 110, 809 P.2d 231 (1991). *See also* Windt, *Insurance Claims & Disputes*, 3d ed., § 11.05 (1995); Ostrager & Newman, *Handbook on Insurance Coverage Disputes*, § 4.02[b][4](9th ed., 1997).

36. *Safeco Title Ins. Co. v. Gannon*, 54 Wn. App. 330, at 334-45, 774 P.2d 30, *rev. denied*, 113 Wn.2d 1026 (1989), *O'Connell, supra*, at 108.

such a report triggers the policy.[37] Policies always include the requirements, however, that the insured give the insurer "prompt" or "immediate" notice. Such requirements can sometimes be harshly applied.[38]

Counsel should also be aware that, unlike most forms of liability insurance, professional liability policies typically provide that expenses incurred in defending the insured will reduce the policy limits. These policies are often referred to as "eroding limits," "cannibalizing" policies, or "wasting" policies. Counsel and the insured should be aware from the outset of a claim whether the policy has this feature, as it will often significantly impact litigation and settlement strategy.

Design professional liability policies also commonly exclude liability for cost estimates and coverage for joint ventures in which the insured participates. Some less restrictive policies will cover the design professional is a joint venture, but only for his or her own negligent acts or omissions.

D. "Wrap-up" Policies

In recent years, some construction projects have been covered by so-called "wrap-up" insurance programs which insure all of the project participants from owner to subcontractors to designers. Depending on the project circumstances, a wrap-up insurance program may be an "owner controlled insurance program" ("OCIP"), or a "contractor controlled insurance program" ("CCIP"). Such programs typically include coverages for general liability, workers

37. *See Concord Hospital v. New Hampshire Med. Malpractice Jt. Underwriting Assoc.*, 137 N.H. 680, 633 A.2d 1384 (1993).

38. *See, e.g.*, Thoracic Cardiovascular Associates, Ltd. v. St. Paul Fire and Marine Ins. Co., 891 P.2d 916 (Ariz., 1995), wherein a complaint was filed against the insured during the "claims made and reported" period, but was not served on the insured until after the policy lapsed. The insurer agreed that the insured had no way to know about the lawsuit until it was served, but the court nevertheless rejected the insured's lack of prejudice and impossibility arguments and granted the insurer's motion for summary judgment.

compensation, builder's risk and excess/umbrella liability; some also include professional liability coverage. The major goal of such controlled or wrap-up insurance programs is to eliminate or minimize problems or disputes that arise all too frequently on major projects as a result of inadequate limits, and restrictive, overlapping or lapses in coverage, indemnity provisions, and problems relating to "additional insured" status.

Project wrap-ups may also be financially advantageous, both in terms of over overall premiums paid and due to more efficient administration of claims. Typically, however, wrap-up insurance programs are implemented only on certain large-scale projects. Ordinarily, each project participant is responsible for obtaining its own insurance coverage, and contracting parties must take appropriate steps to satisfy themselves that those with whom they are contracting have obtained adequate insurance.

II. CONCLUSION

Like killer bees gradually invading, insurance issues are ever-creeping into the construction attorney's practice. Accordingly, a basic understanding of insurance will allow the construction attorney to effectively advise his or her client regarding risk management issues and assist the client in obtaining the benefit of insurance coverages bought and paid for.

Since insurers typically employ counsel well-versed in the applicable policy language and law, it behooves counsel for owners, contractors, and design professionals to also have at least a working knowledge of those matters, or to seek specialized assistance where necessary.

Alternative Dispute Resolution in the 21st Century

13

Alan E. Harris

I. INTRODUCTION

In the days when Perry Mason was a popular television program, lawyers viewed dispute resolution in one of two ways: one side either capitulated (confessed) or a jury decided who would win and who would lose. The methods for resolving disputes are more varied today. The construction field has been in the forefront of the alternative dispute resolution (ADR) movement for many years.

The use of ADR in construction disputes has grown as owners, contractors, subcontractors, designers and sureties have recognized the problems which are sometimes involved in deferring to judges and juries to resolve their disputes. Not only can the court system be expensive, it also has failed in the past to satisfy many of the goals of the industry. A principal perceived inadequacy of the court system was the delay in achieving resolution of disputed issues. Since many entities involved in construction operate on thin profit margins, justice delayed often was justice denied. Court decisions, either at the trial or appellate level, were often of academic interest, but of little value to entities who did not financially survive to benefit from the final decision.

A second problem with the court system was the uncertainty of the result. Judges, despite years of experience, are rarely knowledgeable about construction terms, procedures or standards. Jurors have even less experience; in fact, jurors who might have a background in construction are normally dismissed in the voir dire selection system. While lawyers and experts are paid, and paid handsomely, to educate judges and juries, the extent to which lay people will understand concepts of float, stacking of trades, design versus performance specifications and other construction concepts is unpredictable at best.

The push for alternative means of resolving construction disputes accelerated with the realization that dispute avoidance, or on-site resolution, would assist in achieving a successful job. Lawyers experienced in construction disputes have long recognized that their clients are rarely served by a game of "chicken," where one party threatens to walk off the job and another threatens to use its right to terminate. They, and their clients, now realize that prompt action on disputes (even if only to achieve an interim resolution that will govern conduct until the end of the job) is normally in everyone's best interests.

The techniques currently used to successfully resolve most construction disputes are the subject of this chapter. In increasing resemblance to old-fashioned litigation they are: partnering, dispute review boards, unsupervised negotiation, supervised negotiation, and third-party resolution. Not every technique is appropriate to every project, nor to every dispute. A lawyer who fails to consider the merits of each, however, has failed to adequately serve the client's interests.

II. ON-SITE ADR TECHNIQUES

A. Partnering

Often criticized by those who have not used this technique, partnering is not a New Age method adopted from obscure Eastern religions. Rather, partnering developed within the industry in recognition of the fact that failure of communication is the most common cause of major construction disputes.

Partnering is an attempt to have the principal parties in a construction project agree on common goals before construction begins. These parties meet, usually with a facilitator, for a day or more. The meeting, to be successful, normally takes place at a location where the time can be spent with the minimum of interruptions from the demands of other projects. While every project is different, partnering normally involves key representatives of the owner, general contractor, major subcontractors, designers and others who will have a significant influence on the project. This is one of the few opportunities where parties not in contract with each other (owners and subcontractors, contractors and subconsultants) address each other and deal with important issues.

Increasingly, owners are including partnering meetings as a requirement of the contract documents. Often these provisions will identify the entities to be represented at the partnering meetings, and the categories of people within those organizations who must be present. Successful partnering requires the involvement of field and home office management from all participants, since one of the goals is to establish a number of levels at which the parties can communicate during the project. For example, if issues cannot be resolved in the field, a successful partnering meeting will have established a relationship between the next level of management to whom the issue will be presented, up to and including top management of each of the project participants.

The result of a successful partnering session is frequently the creation of a project charter signed by all participants. These charters are not binding contracts and do not supercede the contract documents. While the charter may contain statements that sound like pap ("a successful and profitable project for all"), the goal is to establish relationships that will see the project through the inevitable disputes.[1]

---

1. Frank Carr et al. Partnering in Construction: A Practical Guide to Project Success (1999); Associated General Contractors of America, Partnering: Changing Attitudes of Construction (1995).

While few attorneys are invited to participate in the partnering sessions (for some reason, lawyers are looked upon as problem creators, not problem solvers), the potential value of this process must be considered in drafting contract documents. At a minimum, attorneys should advise clients on the proper attitude to be taken in these sessions.

B. Dispute Review Boards

This technique recognizes that the input from third parties can assist in the early resolution of disputes. Dispute Review Boards (DRBs or Board) have become extremely popular with the U.S. Army Corps of Engineers, several state departments of transportation and many public entities. DRBs usually consist of three members, one each chosen by the owner and contractor and a third chosen by the two board members. Parties usually pay for the DRB member they have chosen and split the cost of the third member. While attorneys are occasionally selected as DRB members, it is more common for the panel to be composed of people having technical and/or construction expertise.

Normally, the DRB is selected before construction begins and is supplied with copies of the plans and specifications for their use. The Board meets with the owner and contractor on a regular basis so that problems, potential and projected, can be discussed and, if necessary, resolved by the Board. When problems do occur, the DRB receives relevant information, written and oral, meets in executive session and issues a prompt ruling. Because of their experience and their regular involvement in the project, DRB members are able to issue prompt, reasoned decisions.

Depending upon the DRB's charter, its decision may be advisory or binding. Even binding decisions, depending upon the rules established for the DRB, may only bind the parties until completion of the work. At its conclusion, any party dissatisfied with a DRB decision could then challenge the ruling under the dispute mechanism set forth in the contract. Even these interim awards of the DRB, however, have the benefit of keeping the job moving and may effectively resolve the issue if the parties conclude that a later challenge would either be so expensive or problematical as to be unjustified.

While most DRB members remain on the Board throughout the life of the project, case law has developed concerning the right of a party to dismiss its selected member. Obviously unhappy with the decisions the DRB was issuing, the Los Angeles Metropolitan Transit District tried to replace its DRB designee. After lengthy appeals, the District's action was sanctioned. Whether this action so poisoned the process as to render it useless is still unknown.

III. DIRECT, UNSUPERVISED NEGOTIATION

Negotiation is not unique to construction law, and few special rules apply to it. Negotiation is a voluntary process that can occur at any time after a dispute arises, up to and including while the dispute is pending before the appellate courts. Traditional negotiations take place without the assistance of third parties and can involve as many, or as few, people as needed to reach a resolution.

Construction negotiations are frequently more complex because they affect the rights of more than two parties. A successful construction negotiation frequently affects the rights of subcontractors and requires the acquiescence of designers, lenders and others. Rarely is the agreement of two parties adequate for the complete solution to a construction dispute.

Numerous books have been written on the subject and courses on negotiation techniques, tactics and approaches are available from many sources.[2] All stress the importance

2. *See, e.g.*, ROGER FISHER & WILLIAM URY, GETTING TO YES (1981).

of preparation for the negotiation and the thoughtful consideration of the objectives to be achieved. Likewise, the selection of a negotiator with the skills to reach agreement and the knowledge and willingness to deal reasonably with important points as they arise is also crucial. Finally, any negotiated agreement should be immediately reduced to writing to create a binding document reflecting the agreed upon terms while still fresh in the minds of the negotiators.

IV. SUPERVISED NEGOTIATIONS

A. Mediation

Mediation is a non-binding form of negotiation utilizing the assistance of a third-party (the mediator). Although it is increasingly common for contracts to require mediation as a pre-condition to starting a binding form of dispute resolution such as arbitration or litigation, most mediations follow the voluntary agreement of the parties to use this form of ADR. The most important trade organization now requiring mediation in its standard contracts is the American Institute of Architects.[3] Some courts also require the parties to participate in a mediation during the pretrial period. Whether forced mediation is a good, or bad, idea is subject to debate. Those in favor of required mediation argue that any process that forces two parties to meet and consider resolution is valuable. Those who support only voluntary mediation believe that mediation is only successful when both parties agree they are ready to negotiate.

A mediator is a facilitator. The mediator does not decide who has the correct positions or how much money should be exchanged to resolve the dispute. The mediator's role is to make sure the parties understand each other's positions, have

---

3. 3 American Institute of Architects, The Architect's Handbook of Professional Practice: General Conditions for the Contract for Construction § A201 (12th ed. 1994).

their theories tested by probing questions from the mediator, and fully understand the risks and costs of proceeding to the next step in the dispute resolution process. While the mediator does not judge the merits of the cases, a good mediator is much more than the messenger carrying offers from one group to another. An effective mediator will ask tough questions, express skepticism of positions when appropriate, and counsel the parties on the likelihood that settlement positions will assist or detract from the settlement process.

The ideal mediator will have several skills. An ability to listen and the ability to articulate the possible weaknesses in a party's position are probably the two most important attributes of a good mediator. While it is helpful if the mediator is also knowledgeable about the construction process, the mediator's skill in bringing the parties to an agreement is far more important.

Finally, a good mediator will be flexible. A technique that may work with one set of personalities and problems will not necessarily be successful in another context. A mediator must have several speeds; for some parties, it is important that the mediation feel like their "day in court," while for others it is important to "cut to the chase." A good mediator must be able to work in both situations. For all of these reasons, the selection of the mediator may be the most important act a lawyer undertakes in affecting the chances for a successful mediation. Ask for references; talk to parties who have used the mediator; ask about how the mediator would relate to the parties in your dispute; and, if necessary, talk to the prospective mediators to hear first-hand their philosophy of mediation.

The second key step in approaching a mediation is to reach agreement with the other parties on the identity of those who will attend the mediation. Many mediations fail because the people necessary to reach a settlement are not present. This holds true not only for absent insurance representatives, but also for missing corporate officers, sureties and others. It is not enough to be told that the key party is available by telephone. First, it is too easy for someone on the phone to say "no." Additionally, the purpose of the mediation is to exchange positions, judge the merits of those positions and work in a joint manner towards a resolution. The absent decision-maker entirely misses that dynamic.

It is also sometimes important to establish who will not be at the mediation. Since the goal is to reach resolution, someone who is known to be abrasive may be told to stay home. It is also important to decide if experts will be present and to establish timetables for the presentation of positions. All of this can be worked out in advance, with the assistance of the mediator.

The third key activity that should occur before the mediation is preparation. The level of preparation should approach that for a key meeting. A mediation will normally begin with a joint meeting where all parties, their lawyers, their experts, their decision-makers and the mediator are present. Each party is offered the opportunity to make a presentation of its position on each of the matters in dispute. This may be the only opportunity to show the other party's key executives the strength of your case, the weaknesses of the other party's case, the credibility of your witnesses and the skills of your lawyer. The next time everyone meets after an unsuccessful mediation is normally on the courthouse steps on the first day of trial.

Proper mediation preparation includes deciding who will make presentations in the joint session. Occasionally only the lawyers will make statements. In complex construction matters, it is normally beneficial to have presentations by some of the experts and fact witnesses. As mentioned previously, this helps to show the strength and depth of the case. Overheads, Powerpoint presentations, and other forms of visual aids are appropriate and helpful in the joint session. In any presentation, care must be taken to tailor it for the audience—the decision-makers for the other party. The goal of the presentation is to influence those people, not the mediator or the client.

Once each party has made a presentation and the mediator and others have asked questions to clarify positions, the parties are usually put in separate rooms. While they may meet again in other joint sessions later in the mediation, most of the time from this point forward will be in private sessions (commonly called "caucuses"). The mediator will move from party to party discussing factual, legal and other issues. When appropriate, the mediator will carry settlement positions to the other party. In these private sessions, the mediator will encourage the parties to be candid and will guarantee that information disclosed in these private sessions will not be disclosed to the other party without express permission to do so. This candor between parties and the mediator in the private sessions is essential if the mediator is to determine whether there is a realistic chance of a settlement. Failure to be candid with the mediator in the caucus can cause a mediator to abandon hope of a settlement when one, in fact, was achievable.

To promote candor and realistic evaluations, and to minimize posturing, statements made at a mediation are confidential. While many states have statutes that preserve the confidentiality of statements made in mediations, most mediators have everyone present at the mediation sign an agreement confirming this fact. This agreement prevents any party from ever quoting statements made in the mediation at a deposition, hearing or trial or any lawyer from asking a party to repeat what was said at the mediation. Additionally, most states have statutes protecting the mediator from being forced to disclose what was said at the mediation.

Although the parties usually reserve the mediator for a set period of time, a mediation may take less or more time than initially reserved. Sometimes the mediation will be adjourned to permit positions to be reanalyzed or for some key discovery to be taken. In such cases, the mediator will try to get commitments to a timetable for resumption of the mediation. The mediation may go late into the night if the mediator concludes that there is momentum to the mediation that can lead to a successful resolution if the parties keep working. Only as a last resort will the mediator give up and declare the mediation terminated.

For a mediation to be successful, all parties must enter the process with a good faith intention to settle the disputes.

When they reach agreement at the mediation (reports indicate that most cases settle at mediation or shortly thereafter), the parties should immediately reduce their agreement to a writing signed by the necessary officers. Without a signed agreement, the parties may face a second dispute over the terms of the oral settlement agreement.

B. Mini-Trials

Mini-trials are a more structured form of mediation. A facilitator such as a mediator is used; presentations are made to the key decision-makers of both parties; and private negotiation sessions are used after the presentations to negotiate a resolution. As with mediations, mini-trials are normally voluntary, non-binding and confidential. The procedures are flexible and can be tailored to the particular dispute.

A mini-trial is more formal than a mediation in several respects. Normally, the key decision-makers will convene as jurors and sit with the mediator to hear the cases presented. While the attorneys make opening statements as they would in trial, the length of those statements is severely limited. Fact and expert witness offer summary versions of their trial testimony and are subjected to limited cross-examination. The mediator may then offer opinions regarding the merits of the cases and may make predictions of how a judge, jury or arbitrator would resolve the dispute.

After these structured presentations and cross-examinations, the decision-makers/jurors meet privately with the mediator to resolve the dispute. Supporters of this system attribute its success to the empowerment of the decision-makers as evaluators of the merits of the respective positions. This contributes to a business-like approach to resolution, rather than the moneytrader role these decision-makers play in most negotiations. Detractors point out the added expense to the parties in preparing the presentations and the length of time key personnel must spend in preparing for, listening to and negotiating the disputes. All agree, however, that a successful settlement at this stage saves a substantial amount of money over the next step in the dispute resolution process.

V. HYBRID TECHNIQUES—MED-ARB

Med-arb combines the non-binding aspects of mediation with the binding nature of arbitration. Simply stated, the person acting as the mediator changes hats and becomes the arbitrator if the parties are unable to settle their dispute at the mediation. Occasionally, an arbitrator will be asked during the hearings to try to settle the dispute and, if unsuccessful, to resume the arbitration. These processes, which have the support of some trade organizations, rely on the ability of the mediator to change roles and assume the position of a neutral evaluator of the parties' positions.

Supporters of this technique stress its efficiency (because the separate step of selecting an arbitrator is eliminated); its cost effectiveness (since presentations made at the joint mediation session may not need repeating); and its recognition of the reality that good mediators are asked to, and do, evaluate the positions of the parties during the mediation. Opponents note that if the parties know that the mediator will become the arbitrator they will be less candid in the caucus sessions. This, of course, reduces the likelihood that the dispute will be successfully mediated.

Since the selection of the mediator involves an analysis of settlement skills, a proposed mediator/arbitrator's selection must also involve an analysis of his or her adjudicatory skills. Finding one person with all of these attributes may be difficult.

VI. ARBITRATION

A. General Principles

Unlike mediation, arbitration results in a binding, court enforceable decision that, with few exceptions, constitutes the final resolution of the dispute. Absent fraud, conflict of interest or actions that go beyond the scope of the matters that were submitted to the arbitrator, the arbitration award is not reviewed by the state or Federal courts.

Arbitrations are traditionally the product of voluntary agreement by the parties. Frequently, contracts require arbitration and designate the authority under whose auspices the proceedings are to be conducted and the rules to be applied. The most popular purveyor of arbitration services has been the American Arbitration Association which has a large, nationwide group of experienced arbitrators and published sets of rules governing various types of arbitrations, including construction.[4]

Because of abuses in some fields involving consumers, some contractual arbitration provisions have been invalidated. Those clauses usually provided for a biased panel, in locations inconvenient to one party and were included in adhesion contracts. The most criticized contractual provisions have been in the securities, personal loan and medical service areas. Partially in response to these abuses, and partially to make sure that parties are aware that arbitration clauses include a waiver of the right to a jury trial and possibly the right to appeal, some states require the agreement to arbitrate be entered into after the dispute has arisen.

Finally, anyone dealing with an arbitration clause must reconcile it with the provisions of the Federal Arbitration Act. That Act applies to any contract "evidencing a transaction involving commerce," and prevails over any conflicting state law.

B. Contractual Arbitration

The American Institute of Architects, the Associated General Contractors and the Engineers Joint Contract Documents Committee produce separate sets of sample construction contracts. The American Institute of Architects' documents

---

4. American Arbitration Association, Construction Industry Dispute Resolution Procedures (Including Mediation and Arbitration Rules) (1999).

incorporate binding arbitration under the auspices of the American Arbitration Association's Construction Industry Arbitration Rules;[5] the Associated General Contractors' documents offer arbitration as an option to be selected by the parties;[6] The Engineers Joint Contract Documents Committee's forms select litigation as the favored mechanism for final dispute resolution.

The Rules of the American Arbitration Association provide for three separate levels of arbitration depending upon the size and complexity of the case. Small matters involving no more than $75,000 are handled under Fast Track rules that promise a final decision within sixty days of the arbitrator's appointment. These rules put severe limitations on the length of hearings (usually one day) in an effort to provide a prompt, cost-effective resolution.

Cases involving $75,000 to $1,000,000 are handled under the Regular Track rules. These cases are normally heard by one arbitrator. With the exception of document exchanges, pre-hearing discovery is usually not permitted. The arbitrator may suggest limited discovery, especially exchanges of expert reports and depositions of experts as a means of expediting the hearing, but it is rare that the arbitrator will order this over the objections of one party.

Cases over $1,000,000 are handled under the Association's Large, Complex rules. These cases normally have three arbitrators who have the authority to order depositions, interrogatories and other forms of discovery. The Association maintains specialized panels to handle these large cases and the members of these panels have received training in the administration of complex cases.

The introduction of discovery into the arbitration process has not been without controversy. Arbitration was historically sold as a fast, inexpensive means of having disputes

---

5. 3 AMERICAN INSTITUTE OF ARCHITECTS, *supra* note 3.

6. ASSOCIATED GENERAL CONTRACTORS OF AMERICA, ALTERNATIVE DISPUTE RESOLUTION GUIDEBOOK TO AGC 200, AGC 210 (1999).

resolved by knowledgeable construction experts. While this has been retained for Fast Track cases and, to a lesser extent, for Regular Track cases, the American Arbitration Association has determined that the efficient and fair resolution of large cases may require discovery. The parties are not without a remedy if they wish to limit discovery. First, the arbitration agreement can define the types of discovery that can occur; arbitrators will respect those agreements. If the contract is silent, most arbitration rules require that a showing of good cause be presented before discovery will be permitted; in arbitration, discovery is a privilege, not a right.

As with mediation, selection of the arbitrators may be the most important act performed by counsel. By contract, the parties can establish the number of arbitrators who will hear the dispute. They can agree that certain disputes will be heard by a single arbitrator and that others will be heard by three arbitrators. The parties can establish the pre-requisites for the panelists. They can specify years of experience (legal or construction); they can define the composition of the panel (for example, one lawyer, one engineer, one contractor). The contract can specify the method to be used in selecting the arbitrators (for example, from a pre-approved list, from lists proposed by the American Arbitration Association or another agency, from Fellows of the American Institute of Architects or some other honorary society). These contractual limitations will be respected by the authority proposing or selecting the arbitrators.

Once the ground rules for arbitrator selection are established, the attorneys and their clients will normally have limited rights to strike prospective arbitrators without cause and unlimited rights to suggest that prospective arbitrators be removed for justifiable cause. The ultimate decision on challenges to prospective arbitrators will be made by the organization administering the arbitration or by the courts.

Unless incorporated in the arbitration agreement, the rules of evidence will not apply at the arbitration hearing. Arbitrators generally hear any relevant evidence, even if it would be excluded as hearsay under court standards. Rather than excluding the evidence, the arbitrators will evaluate the weight, if any, to be given to hearsay testimony. Likewise, arbitrators may receive evidence by affidavit where court rules would require the presence of the witness at the hearing.

Despite this informality, arbitrators do expect to receive well-organized evidence presented in the most convincing fashion reasonably possible. Where direct testimony could be produced, hearsay testimony will be heavily discounted. Where testimony requires specific expertise, lay testimony on the subject may be received, but will also be discounted.

At the end of the arbitration hearings, the arbitrators will ask for closing statements and/or closing briefs. The parties can arrange for a court reporter to be present at the hearings and the arbitrators may request this for long hearings. Since the hearings may span many days, interrupted by recesses, closing statements and briefs allow the parties to tie their case together and remind the arbitrators of the key evidence. This is helpful to the arbitrators even if they have a transcript available for their use.

If the parties want a reasoned decision (a written award that explains the panel's decision and its calculation of damages), this should be arranged with the arbitrators in advance of the hearings. While some parties are satisfied to get only an award of damages (as they would get from a jury), most parties are interested in understanding the basis for the award. In years past, arbitrators were reluctant to give reasoned awards for fear of reversal by the courts. Modern cases give great deference to arbitrator awards even if the court feels they have inappropriately evaluated the evidence or misinterpreted the case law. As a result of this growing protection of arbitration awards, the reluctance to issue reasoned awards has largely abated.

Most states provide an expedited procedure for having arbitration awards converted into formal judgments. If a party wishes to challenge an award in court, therefore, it must usually act quickly. Each state sets its own rules for the entry of judgments following arbitration awards. While the rules vary from state to state, awards are normally confirmed unless the arbitrators have been guilty of misconduct, conflict of interest or have exceeded the scope of the issues presented to them.

C. Contractual Limitations on the Scope of Arbitration Awards

In drafting the arbitration agreement, the parties can define the limits of the arbitrator's powers. The limitations may define the types of issues that can be decided or the remedies that can be awarded. It is typical for arbitration clauses to limit the types of disputes that can be submitted to arbitration to "any dispute arising out of or related to the contract." A significant body of case law has developed defining the meaning of this type of provision.

More recently, parties have begun to establish limits to the damages arbitrators may award. Some clauses seek to limit awards to damages and prevent the imposition of injunctions or other equitable relief. Increasingly, clauses have limited the amount of damages the arbitrator can award. Two common limitations are the use of "high/low" arbitration or "baseball" arbitration. In high/low cases, the parties agree in advance (but do not normally tell the arbitrator) on the maximum and minimum awards that can be given. If the arbitrator awards an amount outside of those parameters, the award is modified to the nearest amount previously set as the limits.

In baseball arbitration, each party submits a figure to the other (but not to the arbitrator). If neither party accepts the other's figure, the matter goes to arbitration. The arbitrator then issues an award and the party whose figure is closest to the arbitrator's award is the prevailing party and the award is adjusted to coincide with the prevailing party's figure. Both of these techniques are designed to force the parties to realistically evaluate their positions before the arbitration since the prevailing party will be the one who comes closest to the arbitrator's decision.

VII. THE ROLE OF THE COURTS IN THE ADR PROCESS

This chapter has focused on ADR methods selected and fashioned by the parties. The courts, both state and Federal, also have a role in directing the parties into alternative dispute resolution.

For many years, attorneys participated in court-ordered settlement conferences shortly before trial. These conferences were normally conducted by a judge who had set aside an hour or two for the conference and who knew little, or nothing, about the case prior to that time. While this technique helped settle cases that were small and otherwise on the verge of settling, it was not a useful process for larger and more complex cases.

Several United States District Courts began their own arbitration programs in the mid-1980s, and by 1989, Congress began to pass legislation institutionalizing arbitration and other ADR procedures. 28 USC § 652 now requires all district courts to adopt rules offering the litigants at least one form of court-administered ADR and requiring the parties to accept that offer or at least consider some other form of ADR. The Federal Courts have recognized all of the forms of ADR discussed in this chapter and have also added an "early neutral evaluation" process. Under this later system, which takes place early in the life of the case, experienced lawyers volunteer their time to meet with the parties, review key documents and listen to arguments. The evaluator then offers a prediction of the result if the case goes to trial. This opinion is not binding or admissible, but offers the parties a "reality check" at an early stage of their case.

Finally, state courts have also become more sophisticated in using ADR to resolve cases. Since each state, and frequently counties within states, have their own procedures, it is necessary to review local statutes and rules to obtain the ADR procedures. Normally, the full range or techniques are available, whether the arbitrator/mediator is chosen by the court or left to selection by the parties.

Appendix A

Summary of Significant Construction Law Cases

The following are synopses of some of the more important decisions from state and Federal courts involving construction related cases. This is by no means an exhaustive list of significant decisions that have effected the body of construction law as it exists today. There are also cases that are currently before the courts that are being closely followed by practitioners in the area of construction law due to their significance that have not been included in this section.[1] The intent of this appendix is to provide a broad background of past decisions that have shaped construction law.

DAMAGES—"WHO PAYS AND HOW MUCH"

Appeal of Eichleay Corp., CCH 60-2 BCA ¶ 2688 (1960)

General contractor sought recovery of home office expenses allocable to delay in the project caused by the Federal

---

1. Currently one of the most closely-followed cases concerns the constitutionality of race-remedial programs implemented by the government in public contracts. *See* Adarand Constructors, Inc. v. Pena, 515 U.S. 200 (1995) (commonly referred to as "Adarand I"), Adarand Constructors, Inc. v. Pena, 965 F. Supp. 1556 (D. Colo. 1997) (commonly referred to as "Adarand II"), and Adarand Constructors, Inc. v. Slater, 288 F.3d 1147 (10th Cir. 2000) (commonly referred to as "Adarand III"). Since the appeal of issues in this matter are still pending, synopses of these cases have not been included here. The decisions, however, that are issued in these cases clearly affect the way public contracts are issued, and the reader is strongly encouraged to become familiar with this series of *Adarand* cases.

government. At issue was the determination of how to calculate overhead cost, including the home office expenses, that cannot ordinarily be charged to a particular contract, but are nevertheless incurred during and because of the delay caused by the Government.

The Contract Appeals Board recognized that some allocation of home office expenses incurred during the period of delay was an equitable and permitted recovery. The basis of such finding was the recognition that these were expenses that were continuously incurred during the delay, but due to the circumstances of the delay, it was not practical for the contractor to undertake the performance of other work during that period which could have absorbed such costs. Since there is no exact method to determine the amount of such expenses to be allocated to any particular contract or part of a contract, however, the Board accepted the contractor's method allocating the home office overhead costs to specific contracts on a prorated basis, as follows.

The formula accepted by the Board to calculate overhead damages was:

a. $$\frac{\text{Contract Billings}}{\text{Total Billings for contract period}} \times \text{Total Overhead for Contract Period} = \text{Overhead Allocable to the Contract}$$

b. $$\frac{\text{Allocable Overhead}}{\text{Days of Performance}} = \text{Daily Contract Overhead}$$

c. $$\text{Daily Contract Overhead} \times \text{Number of Days of Delay} = \text{Recoverable Overhead Costs}$$

This formula, commonly known as the *Eichleay* formula, has been adopted by many states and Federal jurisdictions as the preferred method of calculating damages for extended overhead costs.

Severin v. United States, 99 Ct. Cl. 435 (1943) *cert. denied* 322 U.S. 733 (1944)

General contractor, on behalf of subcontractor, sought damages from the Federal government which was sustained by the subcontractor due to delay caused by the government, and for the general contractor's overhead costs incurred on account of the subcontractor's delay damages. The court denied the claim to the extent that the claim was actually the subcontractor's claim rather than the actual damages suffered by the general contractor.

The court found that the actual damages for which recovery was being sought were suffered by the subcontractor who had no privity of contract with the government. The court found that the assignment of claim by the subcontractor to the general contractor was forbidden by statute. The court also found that the general contractor was legally protected from any liability to the subcontractor. Consequently, since the general contractor did not suffer actual damages, especially since the general contractor did not pay the subcontractor for its claims, the court held that the general contractor could not recover for damages suffered by its subcontractor.

The Severin doctrine has been limited by many exceptions from later cases. *See Seger v. United States*, 469 F.2d 292 (Ct. Cl. 1972) (Severin doctrine does not apply where the prime contract expressly allows delay damages from government, and the subcontract incorporated the prime contract, such that the subcontractor also has a right of recovery for delay damages). *See also*, *Keydata Corp. v. United States*, 504 F.2d 1115 (Cl. Ct. 1974). Nevertheless, the general principle stated in *Severin* is still followed by most Federal courts.

Foster Wheeler Corp. v. United States, 513 F.2d 588 (Ct. Cl. 1975)

Foster Wheeler Corporation ("FWC") entered into a fixed-price contract with the Philadelphia Naval Shipyard to design, fabricate, and deliver within 13 months two boilers, and perform a "dynamic shock analysis" study that would demonstrate that the boilers could withstand shock up to certain designated intensities set forth in the contract specifications. Since the boilers could not be subjected to actual shock testing, the study was to substitute for the actual testing to determine whether the boiler would be able to withstand a certain amount of shock intensity from external forces. FWC constructed the boilers, and conducted the studies, but the shock intensities demanded by the government were incompatible with the government's additional requirements that the boilers meet certain maintenance, repair and accessibility criteria. Several government reviews and re-tests later, FWC asked for an equitable adjustment to the contract price.

Although the government argued that FWC assumed the contractual responsibility for performing the impossible, the court found that FWC was entitled to an equitable adjustment, based on impossibility. The court adopted the Restatement of Contracts § 454 definition:

> (I)mpossibility means not only strict impossibility but impracticability because of extreme and unreasonable difficulty, expense, injury or loss involved.
>
> "If a contractor proves 'impracticability,' i.e., 'commercial impossibility,' of a contract, it is entitled to recover its costs of attempting to perform the contract, provided that the contractor did not assume the risk of impossibility."
>
> *Id.* at 594 (citations omitted).

Note: Although the Restatement allows relief upon proof of commercial impossibility, the *Foster* court found sufficient evidence of "absolute impossibility" as well. "Absolute impossibility" involves performance "beyond the state of the art."

Kalisch-Jarcho, Inc. v. City of New York, 448 N.E. 2d 413 (N.Y. Ct. App. 1983)

HVAC contractor sought to recover damages it sustained resulting from delays caused by the owner, the City of New York, of 28 months and repeated drawing revisions. The contract contained exculpatory clause which provided that no damages would be awarded in the event of delays. The trial judge instructed the jury that, despite the exculpatory clause, it must find in favor of the contractor if there was "active interference" by the city. The Court of Appeals reversed, stating that there must be more than just active interference to recover in the face of this exculpatory case.

In reversing the trial court, the Court of Appeals held that the no damage clause barred recovery of delay damages. The court held that the no damage clause covered all reasonable and unreasonable delays caused by the City unless the contractor could show that the City acted in bad faith and with deliberate intent.

This case appeared to harshly apply the no damage clause since the clause in question was commonly used and the burden placed on the claimant was high. The court subsequently softened the impact of this case by reaffirming the "not within the contemplation of the parties" exception, the easiest exception to prove to prevent enforcement of the no damage clause." *See Corinno Civetta Const. v. City of New York*, 493 N.E. 2d 905 (N.Y. Ct. App. 1986).

IMPLIED DUTY NOT TO HINDER/IMPAIR—"Stay Out of Our Way"

Luria Brothers & Co. v. United States, 369 F.2d 701 (Ct. Cl. 1966)

Contractor contracted with the Federal government to build an aircraft maintenance facility. During construction, the government's representative became concerned about the bearing capacities of the foundation footings designed to support the concrete columns upon which the concrete roof was to rest. Due to the concern, the government directed the contractor to suspend work while additional subsoil investigations were conducted by the government. The subsoil investigation resulted in several revisions to the foundation plans, as well as the contractor being directed to excavate and perform work in a "stop and go" fashion. The court permitted recovery by the contractor for damages of home office overhead and loss of productivity of labor resulting from improper delays caused by the government.

The basis of the court's decision to permit recovery of damages resulting from delay was its finding that the government had an implied obligation contained in every contract to do nothing that would hinder or delay the other party in performance of the contract. The court found that the original specifications were defective which required changes that were beyond the scope of the original contract, the government was dilatory in recognizing the need for and in making appropriate revisions to the drawings, and that it should not have imposed on the contractor the unreasonable trial and error method of construction. Thereupon, recovery was permitted even though the contractor did not provide evidence as to the amounts of its damages with certainty.

Pearlman v. Reliance Ins. Co., 371 U.S 132 (1962) (White, J., dissenting)

Surety sought to recover remaining contract funds that had been turned over by the government to the bankruptcy trustee of the bankrupt contractor. The surety, pursuant to the terms of the payment and performance bond, completed the contract performance after the contractor had been terminated. The court addressed the issue of who, between the trustee in bankruptcy of a government contractor and the contractor's payment bond surety, had superior right and title to a fund withheld by the government out of earnings due the contractor.

The court held that the surety was equitably subrogated to the rights of laborers and material suppliers, who had been paid by the surety, and had superior rights to the contract funds against the bankruptcy trustees. In so holding, the court found that the surety had an equitable interest in the funds *before* the bankruptcy adjudication, and therefore, the contract funds never became part of the bankruptcy estate.

SPECIAL RELATIONSHIPS—"A BASIS FOR LIABILITY"

Caldwell v. Bechtel, Inc., 631 F.2d 989 (D.C. Cir. 1980)

Caldwell, a heavy equipment operator, allegedly contracted silicosis while he was working in a tunnel under construction for the Washington Metropolitan Area Transit

Authority ("WMATA"). Caldwell was an employee of the general contractor. Bechtel also had a contract with WMATA, as a consultant to provide "safety engineering services" including overseeing the enforcement of safety provisions in relevant safety codes and inspecting job sites for violations.

The District Court dismissed Caldwell's claim against Bechtel, finding no privity and no duty. The D.C. Circuit Court of Appeals, however, reversed the decision based on the principles of duty owed in tort law. The court found that Bechtel owed a duty to take reasonable steps to protect Caldwell from the foreseeable risk of harm caused by excessive silica dust in the tunnels. In so holding, the court determined that by entering into the contract with WMATA, Bechtel undertook the responsibilities of the safety engineer and assumed responsibility for protecting the workers on-site. Bechtel was found to have a "special relationship" with the workers, and the court found that its contract with WMATA gave rise to a tort duty owed to third parties.

The D.C. Circuit later distinguished *Caldwell* in *Wyler v. Korean Airlines Co.*, 928 F.2d 1167 (D.C. Cir. 1991). Plaintiffs in *Wyler* were the survivors of passengers killed when a Korean Airlines jet was shot down by Soviet military. Plaintiffs argued that Air Force personnel who track military planes owed a duty to warn the Federal Aviation Administration ("FAA") that the Korean jet was flying close to Soviet air space. The court found, however, that a duty to protect cannot arise in the absence of some undertaking by defendant to perform some role that relates to the safety of plaintiffs. The court stated that the FAA's undertaking to provide air traffic control services did not mean that Air Force trackers had a special relationship with the passengers to assure their safety, and that the knowledge of one government agency could not be imputed to another government agency. Therefore, the court held that Air Force trackers had no "special relationship" with passengers.

Wood-Hopkins Contracting Co. v. Masonry Contractors, Inc., 235 So. 2d 548 (Fla. Ct. App. 1970)

An action was brought against the subcontractor who performed masonry work on an apartment building. The claims asserted were in connection with water leaks that occurred through the mortar bond between the bricks that were caused by latent defects in the manufacture of the bricks. The bricks that were installed by the subcontractor were bricks that were specified in the contract by their name, Miami Stone, which were manufactured only by the company that bore the same name.

The court held that the subcontractor was not liable for the cost of repairs. In so holding, the court stated that a "builder is not liable for the result of latent defects in material purchased from a reputable dealer." The court found that the subcontractor, acting in good faith, and exercising reasonable care and skill, could not have discovered the latent defect. Additionally, the court found that the contract provision requiring "good quality material, unless otherwise specified" did not give rise to warranty by subcontractor, because specifications called for particular brick by one manufacturer.

Drennan v. Star Paving Co., 333 P.2d 757 (Cal. 1958)

Subcontractor submitted the lowest subcontractor bid to the prime contractor who, relying on the subcontractor bid, computed its overall bid that was then submitted to the owner. The prime contractor was awarded the contract and sought to enter into a subcontract agreement with the subcontractor. The subcontractor informed the prime contractor that it had made a mistake in preparing its bid and refused to honor its bid.

The court found that the prime contractor could recover the difference in the cost between the subcontractor's bid amount and the cost of a replacement subcontractor.

The Court based its decision upon finding that the prime contractor reasonably relied on the bid that was submitted by the subcontractor, and that the bid was an offer which was accepted by the prime contractor upon its reliance on the bid.

In so holding, the court followed the Restatement (Second) of Contracts, Section 87(2). Other courts, however, have refused to find that subcontractor's bids are irrevocable and binding upon the subcontractor. *See Home Elec. Co. of Lenoir, Inc. v. Underdown Heating & Air Conditioning Co.*, 358 S.E. 2d 539 (N.C. Ct. App. 1987).

Home Elec. Co. of Lenoir, Inc. v. Hall and Underdown Heating and Air Conditioning Co., 358 S.E.2d 539 (N.C. Ct. App. 1987)

Contrary to the holding in such cases as *Drennan v. Star Paving Co.*, 333 P.2d 757 (Cal. 1958) (see above), in *Home Elec. Co. of Lenoir*, the court held that North Carolina law prevents a contractor from suing a subcontractor based on the subcontractor's bid, under the theory of promissory estoppel absent a subcontract.

This court considered Section 90 of the Restatement (Second) of Contracts, which some commentators say equates promissory estoppel with a substitute for consideration, but rejected this approach, finding that North Carolina courts have only applied promissory estoppel as a shield to enforce the waiver of a preexisting legal right by a plaintiff. The court further discussed policy reasons against holding a subcontractor liable for its bid including commenting that it would not find the subcontractor's bid irrevocable where the subcontractor, but not the prime contractor, will be bound to the subcontractor's bid.

ARBITRATION—"ISN'T IT MANDATORY?"

Prima Paint Corp. v. Flood & Conklin Mfg. Co., 388 U.S. 395 (1967) (Black, Douglas, and Stewart, JJ., dissenting).

In this case, the United States Supreme Court adopted the divisibility concept to strengthen the enforceability of clauses by making attacks of fraud to invalidate arbitration clauses less effective. The case holds that where a contract contains an arbitration clause, claims of fraud in the inducement of the contract presents an arbitrable issue under the Federal Arbitration Act ("FAA"). The court held that court should decide the issue of fraud in the inducement if the alleged fraud pertains specifically to the arbitration clause, but that the courts cannot issue a decision on the issue of fraud in the inducement of the contract generally.

In so holding, the court adopted the rule previously set by the Second Circuit and found that under Federal law, arbitration clauses are severable from the rest of the contract, and that where no claim is made that fraud was directed to the arbitration clause itself, the broad sweep of the clause includes the claim of fraud in the inducement of the entire contract.

This case has been disagreed with and distinguished by several state and lower Federal courts.

Volt Information Sciences, Inc. v. Board of Trustees of the Leland Stanford Jr. Univ., 489 U.S. 468 (1989) (Brennan and Marshall, JJ., dissenting)

Volt contracted to do electrical work with Stanford. The contract included an arbitration clause. The Federal Arbitration Act ("FAA") was found to be applicable since the contract involved interstate commerce. The contract also had a choice-of-law provision, requiring application of California law. Volt demanded arbitration against Stanford for claims concerning payment. Stanford, in turn, filed a Complaint in Federal court, alleging fraud and breach of contract. Stanford also sought indemnity from two other defendants that did not sign arbitration agreements. Stanford then sought to stay the Volt arbitration, pursuant to a section in California's arbitration act, pending resolution of the litigation. California law was relied upon since the FAA had no comparable section that permits stay of arbitration.

The issue was whether the stay under California law was proper where the FAA was applicable. The court held that because arbitration clauses are contracts, parties are free to structure their arbitration agreements as they see fit. Consequently, the court held that parties to the arbitration agreement could specify that the arbitration agreements be governed by state, rather than Federal law, and that Stanford could properly stay arbitration with Volt, pending resolution of litigation with non-arbitrable defendants.

Additionally, the court held that the Federal Arbitration Act did not preempt the California code permitting the stay. The court stated that the FAA was designed to "overrule the judiciary's long-standing refusal to enforce agreements to arbitrate," and would preempt state laws requiring a judicial forum where parties have an enforceable arbitration agreement. Since the FAA does not prevent parties from structuring the terms of their arbitration agreements, however, and "the interpretation of private contracts is ordinarily a question of state law, which this Court does not sit to review," the court did not find that the FAA preempted the field.

Moses H. Cone Memorial Hospital v. Mercury Constr. Corp., 460 U.S. 1 (1983) (Rehnquist, Burger, and O'Connor, JJ., dissenting)

At issue in this case was a contract for additions to a hospital that contained an arbitration clause. Mercury, the contractor, submitted claims for extra money, which was denied by the hospital, the owner of the project. The hospital filed a declaratory judgment action in state court alleging, inter alia, that Mercury's claims had no legal and/or factual basis and that Mercury had waived its right to arbitration. The hospital also sought indemnity from its architect for any amounts that the hospital might be found liable to Mercury. Mercury filed a motion to compel arbitration in Federal District Court. The District Court ordered a stay of the arbitration, pending resolution of state declaratory relief action. The District Court's decision was reversed by the Court of Appeals, and affirmed by the United States Supreme Court.

The issue was whether Federal courts should abstain from ordering a party to arbitration pending a litigation on its merit in state court. In reaching its decision to enforce the arbitration agreement, the Supreme Court considered the factors that favored abstention by the Federal court from a state litigation: whether the field of law is one peculiarly appropriate for comprehensive treatment in state forums which have the greatest experience and expertise; the absence of any substantial progress in the federal court litigation; and the presence in the suit of extensive rights governed by state law. The Supreme Court followed *Colorado River Water Conserv. Dist. v. United States*, 424 U.S. 800 (1976). None of the *Colorado River* factors operated in favor of abstention.

The hospital contended that the arbitration against Mercury would force it to resolve the case in two different forums since there was no agreement to arbitration between the hospital and the architect. The Supreme Court stated that the purpose of FAA is not necessarily to avoid piecemeal litigation. "Under the Arbitration Act, an arbitration agreement must be enforced notwithstanding the presence of other persons who are parties to the underlying dispute but not to the arbitration agreement." *Moses* at 20.

The Court further stated that the arbitrability of the Mercury/hospital dispute had to be determined by Federal substantive law. The Court recognized that Section 2 of the FAA is a "congressional declaration of a liberal federal policy favoring arbitration agreements, notwithstanding any state substantive or procedural policies to the contrary, and held that the effect of the section is to create a body of federal substantive law of arbitrability, applicable to any arbitration agreement within the coverage of the Act."

KNOWLEDGE AND THE DUTY TO INVESTIGATE

J.A. Jones Constr. Co. v. United States, 390 F.2d 886 (Ct. Cl. 1968)

Plaintiff contractor contracted with Army Corps of Engineers (as agent of the Air Force) to construct a missile test center. The project was part of the Air Force's larger scale Inter-Continental Ballistic Missile (ICBM) program. The contract required premium wages for overtime labor if the contractor fell behind schedule. The ICBM program consumed a great deal of local labor, resulting in a labor shortage. The Air Force did not inform the contractor that its large-scale ICBM program would be going on while the contractor was performing its contract. The contractor and its subcontractors incurred premium labor charges (not contemplated in the contractor's bid) because of the labor shortage, and sought compensation from the Air Force.

The issue in this case was whether the Federal agency had a duty to disclose procurement plans of other Federal agencies which it knew and which might affect the pricing assumptions of bidder.

In this case, the Air Force was responsible for both the ICBM construction project, as well as the contract with the contractor here. The Air Force also had knowledge of the high demand it had placed on the labor in the area, a fact which was not known to the Army Corps, and therefore the Air Force owed the contractor a duty to disclose its plans for a "high priority construction program, premised on the payment of premium wages." *J.A. Jones*, 390 F.2d at 893.

In reaching its decision, the court distinguished its reasoning from that in *Bateson-Stolte, Inc. v. United States*, 172 F. Supp. 454 (1959), where the Atomic Energy Commission was not liable for the Army Corps of Engineers' failure to warn the contractor of the AEC's massive project that consumed much of the local labor. "[A] truly independent federal agency should not be charged with knowledge of what another is doing simply because both are components of the same Federal Government." *J.A. Jones*, 390 F.2d at 891. AEC's project was separate from and unrelated to the contract with the Corps.

Bateson-Stolte, Inc. v. United States, 145 172 F. Supp. 454 (Ct. Cl. 1959)

Plaintiff contractor and the Army Corps of Engineers entered into a contract for construction of a powerhouse.

The Atomic Energy Commission ("AEC") was planning a large-scale construction project that would be performed simultaneously with this contract. Plaintiff contractor had no knowledge about the potential shortage of labor that would be caused by AEC's project and submitted its bid based on current prevailing wage rates. The Army Corps did not disclose to the contractor that the major AEC project could have an impact on the labor availability.

The court held that the contractor could assert a claim against the Army Corps, but remanded the action to trial court to determine whether the Army Corps had knowledge of AEC's project, and whether the contractor had actual knowledge or could have discovered of the AEC's project and its potential for impacting the project if it had exercised reasonable diligence.

SPEARIN AND ITS PROGENY—"YOU GET WHAT YOU PAID FOR"

United States v. Spearin, 248 U.S. 132 (1918)

Spearin contracted with the Federal government to build a dry dock in accordance with the plans and specifications that had been prepared by the government. The work required Spearin to divert and relocate a section of a sewer before the construction of the dry dock could begin. The plans and specifications provided that Spearin should do the work and prescribed the dimensions, material and location of the section to be substituted. Spearin followed the plans and specifications to divert the sewer. However, the plans and specifications provided by the government did not indicate a dam in another sewer that Spearin was directed to connect to. After heavy rain, the dam caused water in the sewer to be diverted, causing internal pressure that broke the sewer that Spearin had relocated and flooded the excavation of the dry dock. Spearin incurred extra costs including delay damages.

The issue was whether Spearin was excused from performance of the contract or entitled to additional compensation because of the unforeseen difficulties encountered.

The court found that generally a contractor that contracts to erect a structure on a particular site assumes the risk of the difficulties that may be encountered at the site. If the contractor is bound to build according to the plans and specifications prepared by the owner, however, the court found that the contractor will not be responsible for the consequences of defects in the plans and specifications. In so holding, the court stated that a contract that prescribes specific performance requirements carries with it an implied warranty that if the plans and specifications are followed, the construction will be adequate. In other words, if a contractor is required to follow the plans and specifications prepared and provided by the owner, the contractor will not be liable for consequences of defects in the plans and specifications.

This case has been distinguished by many lower courts. The basic concept of the *Spearin* doctrine, however, that the owner impliedly warrants the sufficiency of the plans and specifications, continues to be widely accepted.

Trustees of Indiana University v. Aetna Casualty & Surety Co., 920 F.2d 429 (7th Cir. 1990)

Owner brought suit against the contractor's surety for brickwork that had spalled and deteriorated on newly-constructed campus buildings. Due to financial difficulties, the contractor was unable to complete construction and the contract was assigned to the surety who in turn retained another contractor to complete the project. The specifications listed the bricks to be used by manufacturer, trade name and size. Judgment was entered in favor of the surety.

The issue was whether the surety could be held liable for failure of the material that was specified by the owner.

The court upheld the verdict based on the *Spearin* doctrine. In a construction contract, the owner warrants to the contractor that if the contractor follows the owner's plans and specifications, the resulting structure and its component parts will be suitable for the particular purpose intended by the owner. The owner cannot hold the contractor liable for breach of warranty if the damages sustained by the owner are caused because the product expressly selected by the owner or its representative for use on the building was not suitable for such use.

ECONOMIC LOSS DOCTRINE—"Who Can Be Held Responsible?"

J'Aire Corp. v. Gregory, 598 P.2d 60 (Cal. 1979)

J'Aire, a tenant in an airport, leased property from the County to operate a restaurant business. The County contracted with a contractor to renovate the HVAC system in the property. The contract did not specify a time for performance. J'Aire brought an action against the contractor seeking lost profits on the basis that the contractor did not finish the work within a reasonable time.

The court permitted J'Aire's claim for negligent interference with business opportunity due to the contractor's negligence in completing the renovation project within a reasonable time.

In permitting J'Aire to make a claim under a tort theory, the court considered the following factors to determine whether a special relationship between two parties existed, and whether the parties owe a duty of care to each other:

i. the extent to which the transaction was intended to affect the plaintiff;
ii. the foresee ability of harm to the plaintiff;
iii. the degree of certainty that the plaintiff suffered injury;
iv. the closeness of the connection between the defendant's conduct and the injury suffered;
v. the moral blame attached to the defendant's conduct; and
vi. the policy of preventing future harm.

Upon considering the foregoing factors, the court concluded that a contractor owes a duty to complete construction in a manner that will avoid unnecessary injury to the tenant's business, even though the construction contract was entered into with the building owner - not the tenant. The contract was for work that would directly affect the tenant; the risk of harm was therefore foreseeable.

The *J'Aire* decision has received a great deal of negative commentary from other courts. Inasmuch as it permits tort recovery for purely economic losses (in the absence of injury to person or property), it represents a minority rule. *See Coastal Conduit & Ditching, Inc. v. Noram Energy Corp.*, 29 S.W.3d 282 (Tex. Ct. App. 2000).

Florida Power & Light Co. v. Westinghouse Elec. Corp.,
510 So. 2d 899 (Fla. 1987).

This case was decided upon the presentation of a certified question to the Florida Supreme Court from Eleventh Circuit Court of Appeals. Westinghouse contracted with Florida Power & Light to design and manufacture nuclear steam supply systems with steam generators. Florida Power & Light found leaks in the steam generators manufactured by Westinghouse, and brought an action to recover repair costs, alleging breach of warranty claims and negligence.

The court discussed whether Florida Power & Light had a cause of action for recovery under a tort theory. Under Florida law, which follows the majority of courts in other states, the court found that products liability does not give rise to a negligence claim, where there is only "economic loss," i.e., no injury to person or property other than damage to the product itself. Citing *East River Steamship Corp. v. Transamerica Delaval, Inc.*, 476 U.S. 858 (1986), the court stated, "A manufacturer in a commercial relationship has no duty under either a negligence or strict products liability theory to prevent a product from injuring itself."

DEFENSE COSTS—"Who Picks Up The Tab"?

San Diego Naval Federal Credit Union v. Cumis Insurance Society, Inc., 162 Cal.App.3d 358 (1984)

The insured was sued for breach of contract, punitive damages, interference with contract and intentional infliction of emotional distress, and tendered defense of the suit to its insurer. The insurer accepted the tender and retained counsel for the insured who promptly sent a letter of representation to the insured with a reservation of rights by the insurer to disclaim and deny coverage for punitive damages. The insured then retained separate counsel and a dispute transpired regarding whether the insurer was also liable for separate counsel's fees.

The issue discussed in this case was whether the insured is entitled to its own independent counsel to be paid for by the insurer when a conflict exists between the insured and the insurer. The California Court of Appeals held that in situations where there are divergent interests between the insurer and insured as a result of a reservation of rights based on possible non-coverage under the policy, the lawyers hired by the insurer to represent the insured have an obligation to explain the implications of joint representation to the insured. Unless the insured gives informed consent to the dual representation, the insurer must pay the reasonable cost to hire independent counsel to defend the insured, and may not compel the insured to surrender control of the litigation.

Shortly after this decision, the California legislature enacted California Civil Code, Section 2860, which codifies the right to independent counsel where there is certain conflict of interest between the insured and the insurer. Other state legislatures, for the most part, have not gone so far as codifying the right to independent counsel, but many courts in other jurisdictions have recognized the conflict that may exist between the insurer and the insured.

Appendix B
Glossary

Acceleration Claim See also "Constructive Acceleration." A contractor's claim for extra costs resulting from an owner requirement for the contractor to perform and complete its work in a shorter time period than justified by the original contract schedule, plus extensions due to the contractor. Acceleration can be directed or constructive, depending upon the actions of the owner.

Additional Insured Those individuals or entities who generally are not automatically included as insureds under the liability policy of another, but for whom the name insured provides a certain degree of protection under its liability policies. An endorsement is typically required to affect additional insured status for these parties. The named insured's impetus for providing additional insured status to others may be a desire to protect the other party because of a close relationship with that party (e.g., employees or members of the insured club) or to comply with a contractual agreement requiring it to do so (e.g., customers or owners of property leased by the named insured).

Arbitration A referral of a dispute, through mutual agreement or as a result of mandatory contract provisions, to a single arbitrator or a panel of arbitrators, for the rendering of a final and binding determination.

As Planned Schedule The original schedule for a project, reflecting the contractor's intended sequencing and progression of the work.

As-Built Drawings (a.k.a. Record Drawings) Construction drawings revised to show significant changes made during the construction process; usually based on marked-up prints, drawings and other data furnished by the contractor or the architect.

Bar Chart A scheduling document which portrays planned work activities plotted against a time scale.

Bid Protest A rejected bidder's challenge to a public agency's procurement selection process. The right to such a challenge is traditionally derived from statutes and regulations governing public procurement processes.

Bid Bond A form of bid security executed by the bidder as principal and by a surety.

Bid (1)The offer of the bidder submitted in the prescribed manner to furnish all labor, equipment and materials, and to perform the specified work within the time prescribed in the contract, for the consideration of payment at the price or prices stated in the bid schedule. (2) To make a price for work; a proposition, either verbal or written, for doing work and for supplying materials and/or equipment.

Bond See "Bid Bond", "Common Law Bond", "Payment Bond", "Performance Bond", and "Surety Bond".

Builder's Risk Insurance Insurance policy that provides property coverage to reimburse the insured for damage to, or destruction of, a project under construction, regardless of the insured's fault. Generally, builder's risk coverage applies only to buildings under construction, and it lapses or is canceled once the structure is completed and/or beneficially occupied by the owner.

Building Code The legal requirements established by a governing agency creating the minimum requirements for all types of construction within the governing agency's jurisdiction.

Cardinal Change A contractor's claim for breach of contract resulting from an owner-ordered change or changes which so substantially modify the original contract work that the contractor's performance becomes a new and substantially different undertaking.

Change Order A written order to the contractor signed by the owner, and often the engineer or architect, issued after the execution of the contract, authorizing a change in the work or an adjustment in the contract sum or the contract time.

Changed Conditions A contractual provision which allows a contractor to receive additional compensation and an extension of time for conditions which 1) differ materially from those represented to exist in the contract documents, or 2) differ materially from those ordinarily encountered and inherent in the work to be performed under the contract documents.

Commercial General Liability Insurance Commercial general liability insurance (CGL) provides coverage for accidents and injuries to persons or property arising out of the operations of the contractor or other insured. It is intended to cover accidental damage caused by the insured, but not the insured's poor performance of the work called for in the contract. For example, damages resulting from the insured's failure to perform the contract where the only damage is the cost to repair or correct deficiencies in the work are not generally covered. CGL policies provide coverage for damage arising from both bodily injury and property damage, and define these terms to require that the injury result in some actual physical injury or damage. Economic losses and emotional distress damages, without a concurrent physical injury, are not usually covered by a CGL policy.

Common Law Bond Typically, a payment bond whose provisions do not comply with the requirements of a particular statute. Common law bonds have more expanded notice and limitations periods than statutory bonds.

Concurrent Delay The circumstance where a contractor-caused delay to critical path work occurs simultaneously with an owner-caused delay, or an excusable delay.

Consequential Damages Indirect or special damages which do not naturally flow from a breach of contract, but instead, arise from intervening circumstances not ordinarily foreseeable at the time the parties entered into the contract.

Constructability Review A third-party review of project plans and specifications instituted by the owner prior to receipt of contractor bids, made in an effort to identify areas of potential conflict or confusion and for purposes of modifying the design to address such conflicts.

Construction Management (CM) A project delivery system that differs from the traditional design-bid-build system by utilizing the services of a construction manager who becomes a member of the team of owner-architect-construction manager. The project owner's objective in utilizing construction management is often to minimize project time and cost, while maintaining quality, function and aesthetics.

Construction Manager On a project which utilizes a CM delivery system, the construction manager coordinates and communicates the entire project process, utilizing his skill and knowledge of construction to clarify cost and time consequences of design decisions as well as their construction feasibility; the Construction Manager typically manages the bidding, award and construction phases of the project.

Constructive Change A contractor claim for extra compensation, or additional contract time, resulting from the owner's actions which have effectively changed the contract, without following the contract's formal changes provision.

Constructive Acceleration Where a delay to the construction schedule has occurred which would justify an extension to the original project schedule, but the owner refuses to grant the contractor a time extension, forcing the contractor to accelerate its work to meet the unadjusted schedule.

Consulting Engineer A person retained to give expert advice with regard to particular engineering problems.

Contingent Payment Clause See "Pay-When-Paid Clause."

Contract Disputes Act of 1978 Federal statute establishing procedures and requirements for claims submission and dispute resolution of applicable Federal construction claims.

Cost Plus Contract A negotiated agreement between owner and contractor whereby the cost of labor and material are billed to the owner along with the contractor's fee and overhead; fee and overhead is calculated as a percentage of the cost of the work.

Critical Path Method/C.P.M. A network-based scheduling technique whereby a construction project is diagrammed using "fixed" time durations drawn to show the interconnected individual tasks and dependent priorities involved in constructing the project.

Daily Reports Contractor-prepared field reports which track various project conditions, including manpower, equipment and weather, on a daily basis during construction of a project.

Default Nonperformance by a party of its obligations/duties set forth in the contract documents.

Delay or Impact Claim A contractor request for additional compensation or an extension of contract time resulting from an event adversely impacting critical path work which was not the fault of the contractor.

Design Professional The individual who is licensed by a particular jurisdiction, based upon education and training, to prepare plans, elevations, sections and whatever other drawings and specifications may be necessary for the construction of a building or other structure. Architects and engineers, for example.

Design Build Construction contract arrangement in which a single entity (usually a contractor) provides both the design work and building construction services under a single contract with the owner.

Disputes Review Board A panel of neutrals who meet periodically during the course of a construction project to discuss, and, where necessary, render non-binding opinions concerning disputes which have arisen as part of the construction process.

Disruption Claim A contractor's claim for damages resulting from an owner-caused event which increased the contractor's cost of performance without necessarily delaying the overall project.

Early Finish When the date of final completion occurs prior to the amount of contract time stated in an agreement.

Efficiency, Loss of A reduction in a contractor's anticipated rate of productivity, usually resulting from a project delay caused by such things as adverse weather, out-of-sequence work, stacking of trades or overtime work.

Eichleay Formula The most commonly used approach to price a contractor's extended home office overhead costs resulting from an owner-caused delay to critical path work. The formula apportions home office overhead to the contract during actual performance, based on company wide and project specific billings. A daily home office overhead rate is then derived and multiplied by the number of compensable days of delay.

Equitable Lien A claim or lien asserted by an unpaid party upon undisbursed construction funds, usually asserted in the bankruptcy context or pursuant to a "trust fund" statute.

Extra A contractual item of work involving additional contract time or cost to perform.

Federal Acquisition Regulations (FAR) The primary Federal regulations in use by all Federal agencies, governing construction contract procurement and performance.

Final Release A document executed by a contractor relinquishing its rights against an owner in exchange for receipt of final contract payment.

Final Payment Payment made by the owner to the contractor, upon issuance by the architect or engineer of the final certificate for payment, of the entire unpaid balance of the contract sum adjusted by change orders.

Float A comparison of early start or finish dates to late start or finish dates for a particular activity in a network-based schedule.

Flow Down Provision A contract provision which typically transfers risk from a general contractor to a subcontractor by explicitly and specifically incorporating into the subcontract the general contractor clauses and documents for which the contractor wants the subcontractor to assume responsibility.

Force Majeure An event beyond the control of either the contractor or owner which adversely impacts a project's critical path of performance, thereby entitling the contractor to an excusable delay.

Fragnet Short for "fragmentary network"; a partial CPM schedule which contains new schedule activities and logic necessitated by a delay to the project. The fragnet is incorporated into the full updated CPM schedule to determine the overall impact of the delay.

Guaranteed Maximum Price (GMP) Amount established in an agreement between owner and contractor as the maximum cost of performing specified work on the basis of or cost of labor and materials, plus overhead expense and profit.

Home Office Overhead Costs incurred by a contractor's home office necessary to the operation of its business. These costs can become a component of a contractor's claim to an owner when a project's duration extends beyond the contract's completion date and the contractor's home office continues to incur costs in support of the project for a longer period than anticipated at bid time.

Indemnity Clause A contractual provision by which one person or organization agrees to secure another against loss or damage from specified liabilities.

Interference A delaying event to a contractor's performance arising from the owner's hindering the performance without justification.

Jobsite Overhead Commonly referred to as "general conditions," these are all costs necessary for a general contractor to maintain a presence on a project site. These costs typically include such costs as project managers' salaries and fringes, trailers, electricity, telephones and vehicles.

Joint Check Agreement An arrangement whereby a general contractor is authorized to make payments under a subcontract to both the subcontractor and a third party, typically providing labor or materials to a project under the subcontract.

Lien Waiver An instrument by which a person or organization who has or may have a right of mechanic's lien or material lien against the property of another relinquishes such right.

Lien (Mechanic's or Material) A charge put against a project and its underlying real property for satisfaction of unpaid debts on work performed or materials supplied.

Liquidated Damages An amount of money established in a construction contract, usually as a fixed sum per day, as the measure of damages to compensate the owner due to failure of the contractor to complete the work within the contract time.

Liquidation Agreement An agreement between two or more parties who agree to cooperatively align their interests in order to more effectively pursue recovery of their collective damages from a third party.

Lost Labor Productivity Frequent element of construction claims predicated upon the principle that a contractor's workers will accomplish less work if the contractor is prevented from following its planned construction schedule.

Lump Sum Contract Either a negotiated or hard bid agreement whereby the contractor agrees to provide all construction services necessary to complete a construction project, exclusive of change order work, for an all-inclusive price.

Markup The difference between the contractor's cost for labor, materials, subcontractors' fees and equipment, on the one hand, and the amount the contractor bills to the owner. Often expressed as a percentage. For example, the contractor may charge the owner for its labor and materials costs, plus a ten percent markup. The principle behind a "cost-plus" contract.

Master Contract The contract used by a contractor for all of its subcontractors on all of its projects. Specific information regarding a particular project, e.g. price, location, scope of work, is often added to the master contract in the form of a rider or addendum. An efficient alternative to negotiating new subcontracts for each project, a master contract benefits subcontractors, too, by ensuring consistency and predictability.

Multi Prime A project, usually complex, in which the owner proceeds without a general contractor, and contracts separately and directly with each "prime," i.e. major, subcontractor. Often employed in conjunction with an owner/construction manager agreement, to facilitate coordination.

No Damage for Delay Clause A contract provision by which the owner, contractor or subcontractor waives all claims of damage resulting from a particular type of delay in the project. Unenforceable in some states, its main advantage lies in eliminating post-completion home office overhead or consequential damages claims.

Notice to Proceed A written notice to the contractor from the owner directing the contractor to begin working on a construction project. The contract time for completion of the project usually starts to run from the date of the Notice to Proceed.

Overhead, Home Office and Field A contractor's cost of doing business, apart from its immediate labor and materials costs, e.g. utilities. Home office overhead is often not chargeable to a project, while field overhead, such as trailer rental, is typically chargeable as a general conditions cost.

Owner-Controlled Insurance Program (OCIP) Insurance obtained by an owner to cover a large, complex project typically involving many participants. Covers all risks and obviates the need for contractors and subcontractors to obtain their own insurance. It is presumed to promote safety on projects and efficient claims handling.

Partial Release A lien waiver which covers only those payments actually made to the contractor / subcontractor, either to a particular date, or to an amount paid.

Pay If Paid A contract clause obligating a contractor to pay its subcontractors only <u>if</u> it has received payment from the owner. Protects the contractor in the event of owner's non-payment or other default. Often difficult to enforce in some states.

Pay When Paid Similar to the pay-if-paid clause. Limits the contractor's obligation to pay its subcontractors to only *after* it receives payment from the owner. Arguably less restrictive than a pay-if-paid clause, it may still entitle subcontractors to payment in spite of an owner's non-payment.

Payment Bond A mechanism similar to, but by no means the same as, insurance, by which an owner or contractor pays a surety a premium, in return for a guarantee by the surety of prompt payment to subcontractors, if the contractor does not pay the subcontractors.

Performance Bond Similar to a payment bond, but it guarantees an owner that the project will be completed despite contractor default. Required on Federal public works projects under the Miller Act, 40 U.S.C. §§270a *et. seq.*

Prime Contractor A contractor with a particular scope of work, e.g., masonry, which contracts directly with the owner. See "Multi Prime."

Project Architect Term often used to designate the individual or firm which leads the design team for a construction project. May be defined to be a particular person, with particular duties and authority, under the contract between the owner and the contractor.

Project Engineer Term often used to designate a person or firm which designs certain project elements such as mechanical, electrical or structural systems. Often a subcontractor to the Project Architect. Especially on large, complex contracts, may be a joint venturer with the Project Architect.

Project Labor Agreements A universal approach to establishing the rights and responsibilities of trade unions on large, complex, long-running public or publicly-funded projects. Provides certainty for management and labor by, for example, establishing which work falls within the jurisdiction of which trade, and rates to be paid workers on the project.

Program Management A program is the owner's general functional requirements for its project. On large projects, the program is usually a separate, extensive document, cooperatively developed by owner representatives, designers and end users. Program management ensures that program goals will be achieved during and upon completion of construction.

Progress Payment A partial payment to the contractor at arranged milestones of completion. For instance, payments made at 25%, 50% and 100% of completion of the work, or upon completion of major phases, such as foundation and structural steel. Prevents a contractor from financing, or "floating," a project by having to pay subcontractors before receiving payment from the owner.

Project Manager A person or entity who ensures the timely and proper completion of the work, and, oftentimes, timely payment of the contractors or prime contractors, by coordinating the work, managing the paper flow and other components of the project. Sometimes a role played by the Project Architect, but frequently filled by a separately-retained consultant. Seen often on large, complex projects requiring significant coordination of trades.

Project Neutral A third-party arbitrator of disputes, agreed to by the owner and general contractor prior to project commencement. Provides the advantage of contemporaneous alternative dispute resolution. Otherwise, an owner and contractor may have no choice but to stop work until disputes are resolved or to save disagreements until project's end.

Prompt Payment A requirement that the owner pay the contractor within a certain number of days after receiving an application for payment. Can either be pursuant to a contract provision or required by state statute.

Public Works Projects funded by government to provide facilities for public use. Roads, bridges and water treatment plants are types of public works projects. Typically covered by payment and performance bonds, as the property on which these improvements are made is publicly held and therefore not subject to liens.

Punchlist A list of work items remaining to be completed, usually prepared upon substantial completion by the owner, architect and general contractor, following an examination of the site. Final completion is typically defined as occurring when all of the punchlist items are completed.

Purchase Order (P. O.) One of the simplest forms of written contract, oftentimes used to obtain materials.

Retainage A percentage of the contract amount owed to the contractor or subcontractors but held back (i.e. "retained") by the owner to ensure complete performance of the work, especially of punchlist items. Released upon final completion. Sometimes an owner will release retainage early for subcontractors who complete their work early.

RFI ("Request for Information") A formal, written request from the contractor to the architect for clarification of or elaboration on the project plans and/or specifications. Often a precursor to a change order or construction change directive. Can belie, or substantiate, subsequent allegations of unclear or incomplete plans.

Schedule The timetable for project completion. May contain milestone dates, against which progress of the project is compared. Often prepared by the architect, in conjunction with the contractor, based on the owner's required completion date.

Special Conditions Additional contract provisions appearing outside the text of the agreement itself or the contract's general conditions. May clarify, modify or strike standard contractual provisions.

Specifications Verbal design requirements for the project, prepared by the designer and often issued with a request for bid. Provides additional design information with respect to materials, finishes and tolerances.

Stop Notice A written notice in a form required by statute notifying a public owner (or, in some states, a private owner) that a person or entity who has supplied labor, material or equipment to a construction project has not been paid by the party who is to pay them (typically, a notice given by a subcontractor who has not been paid by a general contractor). The effect of the notice is to require the owner to withhold further payments to the contractor until receiving assurance that the subcontractor or other unpaid party has been paid. Failure of the owner to do so may result in the owner having to pay twice.

Subcontractor A specialty contractor or tradesman with whom a general contractor contracts to perform a discrete portion of work, e.g. excavation, grading, mechanical or electrical work. Most project work is performed by subcontractors, otherwise it is known as mechanical work "self-performed" by the general contractor.

Substantial Completion A pre-specified point in the construction of a project, usually when the project is fit to be used for its intended purpose. Any work remaining to be completed is known as punchlist work, the completion of which constitutes final completion.

Sub-Subcontractor An entity with whom a subcontractor contracts to perform a subset of its work. Often a material supplier. For example, a mechanical subcontractor may hire a heating, ventilation and air conditioning (HVAC) sub-subcontractor.

Superintendent A person in charge of the means and methods of a project construction. Typically an employee of the contractor. Oversees the project to ensure completion is in accordance with the plans and specifications.

Suspension When the owner opts to temporarily stop construction, sometimes due to funding problems. Prolonged project suspension may lead to extended job overhead costs, or even demobilization of construction personnel.

T & M or Time and Material A basic method of contracting, by which a contractor is paid a prearranged hourly rate for the time its personnel spend in completing a task, and the expense of the material consumed in that task. Change orders are often performed on a time-and-material basis.

Time Impact Evaluation A means by which the duration and causation of project delays is determined. If a particular stage of construction, such as a concrete foundation pour, is late, a time impact evaluation establishes its effect on subsequent stages which depend upon the delayed event (such as structural steel), and upon the project completion date itself.

Termination Cessation of work, either by the owner or the general contractor, either for cause or without cause. Usually limited to specified events of owner or contractor default; termination can also be at the owner's convenience. If due to owner's default, may trigger contractor's recovery of termination expenses; if due to contractor default, may trigger owner's recovery of additional expense needed to complete the project with another contractor.

Time Extension An extension of the substantial completion or final completion contract dates, due to unforeseen conditions or other delays, the risk of which has not been borne by the owner. Often established by change order. If not, the general contractor may seek an equitable adjustment after project's end, especially if there are liquidated damages, or if the contractor has sustained additional overhead costs.

Unforeseen Conditions Circumstances arising during construction which may require additional time or expense to accommodate. The risk of unforeseen conditions may be borne by the owner or the contractor, depending upon such things as their relative access to information about the site or ability to address the situation, e.g. poor subsurface soil conditions.

Warranty A guarantee provided by a general contractor that the entire work or a particular aspect of it will be free from defects or poor workmanship for a set period of time, often one year. If a defect arises within the time frame, contractor warrants that it will return and repair the work after completion. Typically does not supplant owner's breach of contract remedies, and statutes of repose and limitation related thereto.

Work Authorization The owner's directive to the general contractor to begin certain work. May be either a formal initiation of work pursuant to an executed agreement and general conditions, or may be an informal letter of intent, outlining broad contract terms prior to later execution of a formal written contract.

Appendix C

Useful Websites for the Construction Lawyer

This Appendix provides links to key sources of construction and construction-related law on the World Wide Web. Although some of the web sites listed are commercial in nature, heavy focus is placed on links that make resources available without paid subscription. To facilitate ease of use, web sites are listed under the following major categories:

I. General Sources of Law
   a. Major Portals
   b. Selected Commercial Research Services
   c. Selected Sites Featuring Construction Law Issues
   d. Attorneys and Experts

II. Federal Courts and Regulations
   a. Federal Statutes
   b. Federal Regulations
   c. Federal Rules and Executive Orders
   d. Federal Forms

III. Federal Legislative Resources
   a. Portals & Search Engines
   b. Key Legislative Branch Web Sites
   c. Selected Private Sector Sites Covering Congress

I. GENERAL SOURCES OF LAW

A. Major Portals

AllLaw.com	http://alllaw.com/
Federal Web Locator, The	http://www.infoctr.edu/fwl/
FedLaw	http://www.legal.gsa.gov/
Findlaw.com	http://www.findlaw.com/
Hieros Gamos	http://www.hg.org/
Internet Law Library	http://www.lawresearch.com/

Law.com http://www.law.com/index.html

MegaLaw.com http://www.megalaw.com/

WWW Virtual Library http://www.law.indiana.edu/v-lib/

B. Selected Commercial Research Services

LawGuru http://www.lawguru.com/

Lexis / Nexis http://www.lexis-nexis.com/

VersusLaw http://www.versuslaw.com/

Westlaw http://www.westlaw.com/

C. Selected Sites Featuring Construction Law Issues

American Arbitration Association http://www.adr.org/

American Bar Association
Forum on the Construction Industry
http://www.abanet.org/forums/construction.html

Associated General Contractors http://www.agc.org/

Building Trades Directory http://www.buildingtradesdir.com/

Construction and Design Law Letter (Holland & Knight)
http://www.hklaw.com/

Construction Law Review (Pepper Hamilton LLP)
http://www.constructlaw.com/construct-cfm/currentvolume.cfm

ConstructionRisk.com http://www.constructionrisk.com/

Construction Site Injuries Legal Resource Center
http://www.construction-injuries.com/

FindLaw.com Construction Law Resources
http://guide.lp.findlaw.com/01topics/40construction/index.html

International Conference of Building Officials
http://www.icbo.org/index.html

National Construction Law Center http://www.nclc.com/

D. Attorneys and Experts

ExpertPages.com http://www.expertpages.com/

Lawyers.com (Martindale Hubbell) http://www.lawyers.com/

LawOffice.com (West Group) http://www.lawoffice.com/

Technical Advisory Service for Attorneys (TASA) http://www.tasanet.com/

II. FEDERAL STATUTES AND REGULATIONS

A. Federal Statutes

THOMAS (Library of Congress) http://thomas.loc.gov/home/thomas.html

U.S. Code (Legal Information Institute) http://www4.law.cornell.edu/uscode/

U.S. Code (U.S. House of Representatives) http://uscode.house.gov/usc.htm

B. Federal Regulations

Code of Federal Regulations (NARA) http://www.access.gpo.gov/nara/cfr/

Federal Acquisition Regulations (ARNet) http://www.arnet.gov/far/

Federal Acquisition Regulation (FAR) Supplements (See Links Below)

- Agency for International Development (AIDAR) http://www.usaid.gov/pubs/ads/300/aidar.doc
- Agriculture (AGAR) http://www.usda.gov/procurement/policy/agar.html
- Air Force (AFFARS) http://farsite.hill.af.mil/vfaffar1.htm
- Army (AFARS) http://acqnet.sarda.army.mil/library/zpafar.htm
- Army Corps of Engineers (EFARS) http://www.hq.usace.army.mil/cepr/asp/acquisition/efars.asp
- Defense (DFARS) http://www.acq.osd.mil/dp/dars/dfars.html
- Education (EDAR) http://www.ed.gov/offices/OCFO/contracts/clibrary/edar.html
- Energy (DEAR) http://www.pr.doe.gov/Dear/dear.html
- Labor (DOLAR) http://www.dol.gov/dol/oasam/public/regs/cfr/48cfr/toc_Part2900-2999/Part2900-2999_toc.htm
- Interior (DIAR) http://www.ios.doi.gov/pam/aindex.html
- Justice (JAR) http://www.usdoj.gov/jmd/pss/jarinet.htm#2801

- NASA (NFS) http://www.hq.nasa.gov/office/procurement/regs/nfstoc.htm
- Navy (NAPS) http://www.abm.rda.hq.navy.mil/naps/index.html
- Nuclear Regulatory Commission (NRCAR) http://www.nrc.gov/ADM/CONTRACT/NRCAR.html
- State (DOSAR) http://www.statebuy.gov/dosar/dosartoc.htm
- Transportation (TAR) http://www.dot.gov/ost/m60/tamtar/
- Treasury (DTAR) http://www.treas.gov/procurement/policy/policy.html
- Veterans Affairs (VAAR) http://www.va.gov/oa&mm/vaar/

C. Federal Rules and Executive Orders

Federal Register (NARA) http://www.nara.gov/fedreg/

D. Federal Forms

FedForms http://www.fedforms.gov/

Forms from the Feds (University of Memphis) http://www.lib.memphis.edu/gpo/forms.htm

GSA Forms Library http://www.gsa.gov/forms/

III. FEDERAL LEGISLATIVE RESOURCES

A. Portals & Search Engines

FirstGov.gov http://www.firstgov.gov/

SearchGov.com http://www.searchgov.com/

THOMAS (Library of Congress) http://thomas.loc.gov/

B. Key Legislative Branch Web Sites

U.S. House of Representatives http://www.house.gov/

U.S. Senate http://www.senate.gov/

Congressional Budget Office http://www.cbo.gov/

General Accounting Office http://www.gao.gov/

Library of Congress http://www.loc.gov/

C. Selected Private Sector Sites Covering Congress

ABC News: Congress Watch http://abcnews.go.com/reference/congresswatch/congresswatch_index.html

AOL Government Guide http://www.governmentguide.com/

Congress.org (Capitol Advantage) http://www.congress.org/

Congressional Quarterly http://www.cq.com/home/home.jsp

Roll Call http://www.rollcall.com/

CRS Reports (Penny Hill Press) http://pennyhill.com/

The Hill http://www.hillnews.com/

Washingtonpost.com OnPolitics http://www.washingtonpost.com/wp-dyn/politics/?inside

Yahoo! Politics http://politics.yahoo.com/

IV. FEDERAL COURTS

A. Portals & Search Engines

Courts.net http://www.courts.net/

Federal Court Locator, The http://vls.law.vill.edu/Locator/fedcourt.html

Federal Courts Finder, The http://www.law.emory.edu/FEDCTS/

FedLaw http://www.legal.gsa.gov/

FindLaw Cases and Codes http://www.findlaw.com/casecode/

FirstGov Federal Courts http://firstgov.gov/us_gov/judicial_branch.html

B. Federal Courts

U.S. Supreme Court http://www.supremecourtus.gov/

U.S. Court of Appeals

- Federal Circuit http://www.fedcir.gov/
- District of Columbia Circuit http://www.cadc.uscourts.gov/

- First Circuit http://www.ca1.uscourts.gov/
- Second Circuit https://www.tourolaw.edu/2ndcircuit/
- Third Circuit http://www.ca3.uscourts.gov/
- Fourth Circuit http://www.ca4.uscourts.gov/
- Fifth Circuit http://www.ca5.uscourts.gov/
- Sixth Circuit http://www.law.emory.edu/6circuit/
- Seventh Circuit http://www.ca7.uscourts.gov/
- Eighth Circuit http://www.ca8.uscourts.gov/
- Ninth Circuit http://www.ca9.uscourts.gov/
- Tenth Circuit http://www.ck10.uscourts.gov/
- Eleventh Circuit http://www.ca11.uscourts.gov/opinions.htm

U.S. Court of Federal Claims
http://www.law.gwu.edu/fedcl/opinions.htm

U.S. Court of International Trade http://www.uscit.gov/

U.S. District and Bankruptcy Courts
http://www.fedaccess.com/district-bankruptcy-courts.htm

V. FEDERAL GOVERNMENT AGENCIES

A. Portals & Search Engines

AOL Government Guide (America Online)
http://www.governmentguide.com/

Acquisition Reform Network (ARNet) http://www.arnet.gov/

Federal Web Locator http://www.infoctr.edu/fwl/

FirstGov Government Agency Locator http://www.firstgov.gov/

U.S. Blue Pages http://www.bluepages.gov/index.cfm

B. Boards of Contract Appeals (BCAs)

Armed Services BCA http://www.law.gwu.edu/asbca/decis00.htm

Army Corps of Engineers BCA http://www.usace.army.mil/ENGBCA/

General Services Administration BCA http://www.gsbca.gsa.gov/

Government Printing Office BCA
http://www.contracts.ogc.doc.gov/gpobca/

Federal Aviation Administration ODRA http://www.faa.gov/agc/

Labor BCA http://www.oalj.dol.gov/libbca.htm

U.S. Department of Agriculture BCA http://www.usda.gov/bca/

Veterans Affairs BCA http://www.va.gov/bca/index.htm

C. Other Federal Agency Review Forums

Comptroller General (General Accounting Office)
http://www.gao.gov/decisions/decision.htm

Department of Labor Administrative Review Board
http://www.dol.gov/dol/arb/welcome.html

Department of Labor Benefits Review Board
http://www.dol.gov/dol/brb/welcome.html

National Labor Relations Board http://www.nlrb.gov/

Occupational Safety and Health Review Commission
http://www.oshrc.gov/

Small Business Administration Office of Hearings and Appeals
http://www.sba.gov/oha/

United States Postal Service Purchasing Protest Decisions
http://www.usps.gov/lawdept/LegalPly/Contract/

D. Other Selected Federal Government Sites of Interest to the Construction Community

ACQWeb (DoD Acquisition Web Site) http://www.acq.osd.mil/

Energy Efficiency and Renewable Energy Network (EREN)–Department of Energy http://www.eren.doe.gov/

Federal Procurement Data System (GSA)
http://fpds.gsa.gov/fpds/fpds.htm

FedStats http://www.fedstats.gov/

Environmental Protection Agency Office of Solid Waste
http://www.epa.gov/epaoswer/osw/mission.htm

Environmental Protection Agency Superfund Web Site
http://www.epa.gov/superfund/index.htm

National Technical Information Service http://www.ntis.gov/

U.S. Census Bureau - Construction Statistics
http://www.census.gov/const/www/index.html

VI. STATE AND LOCAL RESOURCES

A. Portals & Search Engines

Courts.net http://www.courts.net/

FedAccess.com
Links to State Government Web Sites
http://www.fedaccess.com/govtwebsites.htm

FindLaw.com State Resources http://guide.lp.findlaw.com/11stategov/

Internet Law Library - U.S. State and Territorial Laws
http://www.priweb.com/internetlawlib/17.htm

LawDog.com State Laws and Agencies
http://www.lawdog.com/state/laws.htm

LawStreet.com http://www.lawstreet.com/

Legal Information Institute
State Statutes (Cornell University)
http://www.law.cornell.edu/statutes.html

Library of Congress State and Local Resources
http://lcweb.loc.gov/global/state/

State and Local Government on the Net (Piper Resources)
http://www.piperinfo.com/state/index.cfm

VII. NON-PROFIT ORGANIZATIONS AND TRADE ASSOCIATIONS

A. Portals & Search Engines

AssociationCentral.com http://www.associationcentral.com/

ContractorGuide.com http://www.contractorguide.com/Associations/

FedLaw Professional Associations
http://www.legal.gsa.gov/legal91.htm

B. Selected Non-Profit Organizations and Trade Associations

American Arbitration Association http://www.adr.org/

American Association of State Highway Transportation Officials (AASHTO) http://www.transportation.org/aashto/home.nsf/FrontPage

American Bankruptcy Institute (ABI World) http://www.abiworld.org/

American College of Construction Lawyers http://www.accl.org/

American Concrete Institute http://www.aci-int.org/

American Construction Inspectors Association http://www.acia-rci.org/

American Institute of Architects http://www.aiaonline.com/

American Institute of Constructors http://aicnet.org/

American Institute of Steel Construction http://www.aisc.org/

American Institute of Timber Construction http://www.aitc-glulam.org/

American Law Institute http://www.ali.org/

American Road and Transportation Builders Association http://www.artba.org/

American Society of Civil Engineers http://www.asce.org/

American Society of Heating, Refrigerating and Air-Conditioning Engineers, Inc. http://www.ashrae.org/

Associated Builders and Contractors http://www.abc.org/

Associated Equipment Distributors http://www.aednet.org/

Associated General Contractors http://www.agc.org/

Associated Owners and Developers http://www.constructionchannel.net/profdirectory/html/associated_owners___developers.html

Associated Schools of Construction http://www.ascweb.org/

Association for Transportation Law, Logistics and Policy http://www.transportlink.com/atllp/

American Subcontractors Association http://www.asaonline.com/

Design-Build Institute of America http://www.dbia.org/

Civil Engineering Research Foundation http://www.cerf.org/indexjs.htm

Construction Financial Management Association http://www.cfma.org/

Construction Industry Institute http://www.construction-institute.org/

Construction Industry Manufacturers Association http://www.cimanet.com/

Construction Industry Roundtable, Inc. http://www.cirt.org/

Construction Management Association of America http://www.cmaanet.org/

Construction Marketing Research Council http://cmrc.net/

Construction Owners Association of America http://www.coaa.org/

Construction Safety Council http://buildsafe.org/

Construction Specifications Institute http://www.csinet.org/

Construction Writers Association http://www.constructionwriters.org/

Defense Research Institute http://www.dri.org/

Design-Build Institute of America http://www.dbia.org/

Design-Build Manufacturers Association http://www.dcenter.com/

Energy and Environmental Building Association http://www.eeba.org/

Equipment Manufacturers Institute http://www.emi.org/

Federal Facilities Council http://www4.nas.edu/cets/ffc.nsf

Federation of Insurance and Corporate Counsel http://www.thefederation.org/

Institute for Electrical and Electronics Engineers http://www.ieee.org/

International Code Council http://www.intlcode.org/

International Facility Management Association http://www.ifma.org/

International Silo Association http://www.silo.org/

Mason Contractors Association http://www.mcastle.com/

Masonry Society http://masonrysociety.org

Materials Technology Institute of the Chemical Process Industries
http://www.mti-link.org/

Metal Building Manufacturers Association http://www.mbma.com/

Metal Construction Association http://www.mca1.org/

National Association of Corrosion Engineers http://nace.org/

National Association of Home Builders http://www.nahb.com/

National Association of Minority Contractors
http://www.n-a-m-c.com/

National Association of Women in Construction
http://www.nawic.org/

National Association of Surety Bond Producers
http://www.nasbp.org/

National Contract Management Association http://www.ncmahq.org/

National Electrical Contractors Association http://www.necanet.org/

National Frame Builders Association http://www.postframe.org/

National Housing and Rehabilitation Association
http://www.housingonline.com/

National Railroad Construction and Maintenance Association, Inc.
http://www.nrcma.org/

Power and Communication Contractors Association
http://www.pccaweb.org/

Professional Construction Estimators Association
http://www.pcea.org/

Steel Erectors Association of America http://www.seaa.net/

Systems Builders Association http://www.systemsbuilders.org/

Surety Association of America http://www.surety.org/

Women Construction Owners and Executives
http://www.wcoeusa.org/

VIII. SELECTED CROSS-DISCIPLINARY SITES FEATURING CONSTRUCTION LAW ISSUES

A. Government Contract Law

American Bar Association Section of Public Contract Law
http://www.abanet.org/contract/home.html

Federal Access http://www.fedaccess.com/

Fedmarket.com http://www.fedmarket.com/

GovCon.com http://www.govcon.com/

Professor Steven Schooner's Government Contract Law Site
http://www.law.gwu.edu/facweb/sschooner/

Where in Federal Contracting? http://www.wifcon.com/

B. Insurance Law

American Bar Association Torts and Insurance Practice Section
http://www.abanet.org/tips/home.html

ConstructionRisk.com http://www.constructionrisk.com/

Mealeys.com http://www.mealeys.com/

C. Transportation Law

American Association of State Highway Transportation Officials (AASHTO)
http://www.transportation.org/aashto/home.nsf/

American Road and Transportation Builders Association
http://www.artba.org/

Association for Transportation Law, Logistics and Policy
http://www.transportlink.com/atllp/

LawGuru.com Transportation and Admiralty Law
http://www.lawguru.com/ilawlib/111.htm

Transportation Library (Northwestern University Library)
http://www.library.northwestern.edu/transportation/

IX. MISCELLANEOUS BUSINESS RESEARCH TOOLS

A. Selected Web Sites

Construction.com (McGraw Hill)	http://www.construction.com/
Construction Notice Services, Inc.	http://cnslien.com/
Construction Site, The	http://www.constr.com/tcs.htm
EDGAR-Online (Corporate SEC Filings)	http://www.edgar-online.com/
Internet Tools for Lawyers	http://www.netlawtools.com/
MapQuest	http://www.mapquest.com/
National Lien and Bond Claim Systems	http://www.mechanicslien.com/
RefDesk.com	http://www.refdesk.com/
Salary.com	http://www.salary.com/
State Permits, Inc.	http://www.statepermit.com/
The Blue Book of Building and Construction	http://www.thebluebook.com/
Witness Systems	http://www.witness.com/
Yahoo! Maps	http://maps.yahoo.com/

X. ONLINE NEWS RESOURCES

A. Selected Construction News Sources

AEC Automation Newsletter	http://www.aecnews.com/
Architectural Record	http://www.architecturalrecord.com/
Builder Online	http://www.builderonline.com/
Buildings Magazine	http://www.buildingsmag.com/
ConstrucTECH	http://www.constructech.com/
ENR.com (Engineering News Record)	http://www.enr.com/
Environmental Design & Construction Online	http://www.edcmag.com/

F.W. Dodge http://www.fwdodge.com/

National Home Center News http://www.homecenternews.com/

RSMeans.com http://www.rsmeans.com/

Yahoo! Construction/Building News
http://biz.yahoo.com/news/construction.html

XI. SELECTED E-COMMERCE WEB SITES

A. Electronic Procurement

Bidline.com http://www.bidline.com/

CBDNet (Commerce Business Daily Online)
http://cbdnet.access.gpo.gov/

FedBizOpps.gov http://www.fedbizopps.gov/

ProcureZone.com http://www.procurezone.com/

B. Electronic Project Collaboration / Management

Accela http://www.accela.com/

Advanced Rendering Technology http://artrender.intelligentorgs.com/

Arcom Master Systems (Masterspec) http://www.arcomnet.com/

Autodesk http://www.autodesk.com

BuildSoft http://www.buildsoft.com.au/

Buzzsaw.com http://www.buzzsaw.com/

Cadence http://www.altagroup.com/

Constructware.com http://www.constructware.com/

Cyberserv, Inc. http://www.cyberserv.com/

Deltek Systems http://www.deltek.com/

J.D. Edwards http://www.jdedwards.com/

Meridian Project Systems http://www.mps.com/

MP Interactive http://www.e-builder.net/index2.html

Omware, Inc. http://www.omware.com/

ThePowerTool.com	http://www.thepowertool.com/
Timberline Software	http://www.timberline.com/
Viecon.com (by Bentley)	http://www.viecon.com/
VisionPlanner.com	http://www.visionplanner.com/

C. Online Vendors and Suppliers

BuildFind Network	http://www.buildfind.com/
Dirtpile.com	http://www.dirtpile.com/
GovCon (Government Contracts Portal)	http://www.govcon.com/
Sweet's Product Marketplace	http://www.sweets.com/

Index